PRAISE FOR
FUSION CAPITALISM

"Steve's book is a fresh and inspirational take on how conservative values and clean energy are perfectly aligned. The tired, old arguments of the past that clean energy costs too much or does not work have been long refuted. It's time for conservatives to rally around a US commitment to win the war on climate change."

—**Jigar Shah**, *CEO, Generate Capital*

"*Fusion Capitalism* should be in the hands of every governor, US senator, House representative, and state lawmaker in America. Climate change is fast approaching the point of no return, and clean energy is the proven, practical, and profitable solution. Let's reach across the aisle and work together as a human family to not only save the world but make it a healthier and more prosperous one for future generations."

—**Ted Strickland**, *Governor, State of Ohio (2007–2011)*

"Steve's heartfelt passion for growing the tent of bipartisan advocates for the clean energy economy and sustainability is deeply inspiring. The collective future of our nation and our world depends on everyone embracing long-term growth and prosperity over short-term interests. Steve's commitment to leaving the world a better place for future generations is evident by both his actions as well as his leadership as the founder and CEO of the Melink Corporation. I am grateful for Steve's influence in this movement and for modeling true market transformation for all of us."

—**Mahesh Ramanujam**, *President and CEO, USGBC & GBCI*

"In addition to building a nationally recognized business helping commercial clients improve their profitability through energy efficiency and renewable energy, Steve Melink has been a true civic leader across the industrial Midwest heartland in advocating the economic and wealth-creation opportunities associated with accelerating the transition to a clean energy economy. With *Fusion Capitalism*, Steve illuminates how to integrate strong business acumen with a values-based philosophy to create a better future for customers, employees, owners, and society."

—**Richard T. Stuebi**, *President, Future Energy Advisors*

"Steve brings his real-life storytelling style and passion for the environment to a critical issue for all of us—how do we carve a path to the future in between the traditional stereotype of right-wing fiscally conservative anti-environment republicans and left wing big government pro-environment liberal democrats? The truth, and the right course of action for our world, is always somewhere in between. A must-read!"

—**Tom Kirkpatrick**, *President and CEO, Eco Engineering, Inc.*

"Steve has been my mentor and friend for nearly fifteen years. He's taught me that business should serve a greater good beyond just profits. In *Fusion Capitalism*, Steve provides a framework for solving humanity's greatest problem and opportunity—climate change. Therefore, it's a must-read for conservatives, especially those who are 'challenged' on this issue. This matter is not partisan—it's a moral imperative upon which we are reaching a tipping point. Can Americans on both sides of the aisle rise to the challenge? Can we lead the clean energy revolution? Read this book and learn from one of the best."

—**Craig Davis**, *President, Melink Corporation*

"For Christian businessman Steve Melink, it's all about the three Ps: people, profit, planet. In his new book, *Fusion Capitalism*, he advocates for a clean energy revolution grounded in principles he learned growing up in a conservative household in the suburbs of Cincinnati. Like Teddy Roosevelt, Melink ties his commitment to protecting the natural world to the true meaning of conservatism. Embracing a clean energy future is 'completely in line with American principles, family values, and conservative ethics,' he argues. As the major political parties battle over climate change and energy policy, *Fusion Capitalism* shows conservatives how to create jobs, improve public health, drive innovation, and preserve our planet for future generations—an agenda that could give the Republican Party a competitive future."

—**Brewster Rhoads**, *CEO, Brewster Rhoads & Assoc*

"In *Fusion Capitalism*, Steve intertwines a deeply personal account of his life experience with wisdom gained through a successful professional career. The central argument—that environmentalism and conservatism are inextricably linked—is a message that needs to be heard. For too long we have allowed politics and partisanship to dictate our views on climate change, when in fact addressing these challenges will require the kind of vision that policy makers rarely possess. In laying out 'A Conservative Platform for America's Clean Energy Future' Steve not only draws on the key elements of his argument; he also creates an ambitious, yet achievable, framework for moving forward. *Fusion Capitalism* is a book that will enlighten and inspire people across the ideological spectrum to take up the cause of a clean energy future."

—**Jane Harf**, *Executive Director, Green Energy Ohio*

"This book makes a compelling business case for a zero-energy future in the building industry. We don't have to wait until 2030 or 2040; we can start doing this now. The technologies and best practices are available and cost effective. In fact, Steve has shown at his new HQ2 that the cost premium to go zero-energy was less than 10 percent over a conventional building. Given this, every home and building owner in America should be thinking about this opportunity. Our company was proud to work side by side with Steve to help advance his vision for a brighter future. He's a true pioneer and leader. And his building is a model for the world to see."

—**Kevin Scott**, *President, Bunnell Hill-Schueler Group*

"Steve Melink lays out a vision based on his traditionally conservative midwestern values: faith, family, community, entrepreneurial passion, and stewardship. His own fusion of these values creates a compelling argument for a conservative embrace of renewable energy and energy efficiency. *Fusion Capitalism* provides both the rationale and the plan to take on the challenge of climate change."

—**Mark Shanahan**, *CEO, New Morning Energy*

"Steve is a pioneer and visionary in Ohio's business community. He was one of the first to see the clean energy revolution coming, and instead of just joining it, he decided to help lead it. He has enjoyed great commercial success as a result but more importantly become a respected thought leader across our state and throughout the quickly changing energy sector. *Fusion Capitalism* should be required reading for every government and business leader in the country."

—**Terrence N. O'Donnell**, *Chair, Government Relations, Dickenson Wright*

"Ethical and effective leaders marshal their resources to grow the triple bottom line: people, profit, and planet. In *Fusion Capitalism* Steve offers a rational prescription and call for action to conservative leaders. He argues (successfully) that it is time for stewards of organizations to maximize the good for all stakeholders, not just shareholders. Hey, leaders, the sun is shining ... it's time!"

—**Bob Pautke**, *President, SOAR with Purpose*

"Steve Melink is a practical, conservative businessman—a leader who understands what the future looks like. For over a decade Steve and his company have helped the Cincinnati Zoo grow our efficiency and sustainability program, enabling us to become recognized as 'the greenest zoo in America.' *Fusion Capitalism* makes a business case for clean energy that even Fox News watchers will embrace."

—**Thane Maynard**, *Executive Director, Cincinnati Zoo & Botanical Garden*

"Being a conservative shouldn't mean you aren't a conservationist. The root word 'conserve' and its meaning is the same. Reagan and Thatcher patched the ozone hole by leading the world to ban chlorofluorocarbons. Steve Melink articulates a clear vision that clean energy is about innovation and leadership, not Luddism or liberalism. Steve inspires today's businesses to lead the next energy revolution by harnessing the sun, our most abundant and ultimately most cost-effective source of power."

—**Brian Ross**, *Principal, Mid-Market Growth Partners*

FUSION CAPITALISM

STEVE MELINK

FUSION CAPITALISM

A CLEAN ENERGY VISION FOR CONSERVATIVES

ForbesBooks

Published by ForbesBooks, Charleston, South Carolina.
Member of Advantage Media Group.

ForbesBooks is a registered trademark, and the ForbesBooks colophon is a trademark of Forbes Media, LLC.

Printed in the United States of America.

10 9 8 7 6 5 4 3 2 1

ISBN: 978-1-950863-21-1
LCCN: 2020914337

Cover design by David Taylor.
Layout design by Carly Blake.

This custom publication is intended to provide accurate information and the opinions of the author in regard to the subject matter covered. It is sold with the understanding that the publisher, Advantage|ForbesBooks, is not engaged in rendering legal, financial, or professional services of any kind. If legal advice or other expert assistance is required, the reader is advised to seek the services of a competent professional.

Advantage Media Group is proud to be a part of the Tree Neutral® program. Tree Neutral offsets the number of trees consumed in the production and printing of this book by taking proactive steps such as planting trees in direct proportion to the number of trees used to print books. To learn more about Tree Neutral, please visit **www.treeneutral.com**.

Since 1917, Forbes has remained steadfast in its mission to serve as the defining voice of entrepreneurial capitalism. ForbesBooks, launched in 2016 through a partnership with Advantage Media Group, furthers that aim by helping business and thought leaders bring their stories, passion, and knowledge to the forefront in custom books. Opinions expressed by ForbesBooks authors are their own. To be considered for publication, please visit **www.forbesbooks.com**.

This book is dedicated to all the clean energy entrepreneurs, innovators, and early adopters who took significant risks to advance a cause greater than themselves. Though victory is yet to be won, the path forward is clear, and you have bequeathed a brighter tomorrow to all humanity.

CONTENTS

America, America, God Shed His Grace on Thee.
And Crown Thy Good with Brotherhood.
From Sea to Shining Sea.
—KATHERINE LEE BATES, 1895

Al Gore Invented the Environment

O n any given day, I drive into Madeira to pick up some groceries, buy a few things at the pharmacy, or browse for parts at the hardware store. I live with my wife, Mary Frances, in Indian Hill, a small village northeast of Cincinnati. It's an idyllic midwestern community of 6,900 residents with beautiful homes and lawns, large shade trees dotting the landscape, and a brick firehouse and ranger station at the main intersection.

You can probably imagine the soft flow of traffic moving up and down the streets of this traditional bedroom community—the school buses picking up and dropping off children on weekdays, and everyone getting dressed up for church on Sunday. It reminds me of the quaint town I grew up in not far away. I appreciate the green spaces, wholesome culture, and family values here.

Clearly, I'm no revolutionary. But I'm going to let you in on a secret: I love driving my all-electric Tesla down Miami Avenue and showcasing the future. It's not a status symbol; I just want to do my part to help promote the next great technology wave and hopefully inspire others to make the leap too. Electric cars are still a relative rarity in this area, but make no mistake: they're the way of the future. And once you drive one—well, it's all but impossible to go back to conventional gas-fueled cars.

I know some of my neighbors shake their heads, believing electric cars are a fringe, environmentalist curiosity—not a "real car." I know they probably think the same of the solar panels installed on the side of my house. And even though the word *environmentalist* should be a compliment, they don't intend it that way. That's okay; they'll eventually come around … because they'll have to.

In fact, clean energy technologies are not some passing fad. They're quickly going mainstream, and we can see this more and more every day. Just as smartphones have disrupted the telecom industry, e-commerce has disrupted brick-and-mortar retail, and so many other innovations have turned previously successful models into dinosaurs—clean energy technologies are poised to forever change the world, turning fossil fuel power plants and refineries into fossils themselves.

Having said this, I have to admit it's confusing and even frustrating that the conservative brand that I grew up and identified with for most of my life is dismissing much of the damage that coal, oil, and natural gas are doing to our health and environment—and missing out on the clean energy economy of the future. Why? I often wonder. Conservatives never fled aviation, space exploration, medical advances, computers, or the internet age. The power of innovation and free markets has always been a core pillar of conservative politics. So why have the clean energy movement and all its benefits and promise become partisan issues and often resisted tooth and nail by the American right? Did it start when Al Gore released his climate change documentary, *An Inconvenient Truth*, in 2006 and then become doctrine when Barack Obama supported clean energy during his presidency?

Given the old punch line that "Al Gore invented the internet," it's ironic that fellow conservatives have essentially credited him with inventing climate change science. Conservatives have long promoted the philosophy that we protect the environment for our children and future generations. Planet Earth is a speck in a vast universe with billions of galaxies and the only place we know of that supports life. Surely it is one of God's greatest gifts to us as human beings. Therefore, it is a practical and moral imperative that we give our planet the same respect that many of our conservative forebears did.

Teddy Roosevelt set aside millions of acres for national parks and wildlife. Dwight Eisenhower expanded the national park system, protecting forests, bird and game reserves, and fisheries. Richard Nixon created the Environmental Protection Agency (EPA) and signed the Endangered Species Act. George H. W. Bush strengthened the Clean Air Act and was the first president to identify climate change as a major federal priority. And during his presidency, George W. Bush signed legislation to promote solar and wind power as well as higher lighting efficiency standards. Caring for America's natural resources and taking action to counter climate change have not been, and should not now be, the purview of one party.

I grew up in a conservative household and attended Catholic schools as a child. One of the most important ethics taught to us was to respect Mother Nature. We could all do simple things like not litter, pollute, and prevent forest fires. This was back when dirty air, land, and water were becoming increasingly endemic across America. Smog was blanketing our cities, acid rain was destroying our forests, and ozone depletion was causing a hole to develop in Earth's atmosphere.

Through educational awareness programs as well as smart government policies, we successfully addressed these life-threatening problems over time. Our homes, communities, and country were getting safer and healthier again. Economic growth and environmentalism could coexist and, in fact, complement each other and yield a better total outcome. Clean air, water, and land was not a conservative issue or progressive issue. It was a human issue.

Enter the year 2000 and a closely contested presidential election. Al Gore had been vice president for eight years in the previous administration, and when he lost to George W. Bush, he retreated to what he knew best. Though many of his earlier positions were fair game for criticism from those on his right, it should be acknowledged that the emerging science of climate change was something he was studying when few others were. The reflexive muscle that caused many Republicans to oppose him and his views on this science came from the fossil fuel lobby and emotional partisanship, not the public interest or intellectual honesty.

> **I'd like to make the case that the science of climate change and the extraordinary opportunities of the clean energy economy are entirely in keeping with conservative values.**

In this book I'd like to make the case for conservatives to rethink their position on climate change, as I have. And rather than simply adopting the talking points of the left, I'd like to make the case that the science of climate change and the extraordinary opportunities of the clean energy economy are entirely in keeping with conservative values.

Time out, you say. Can we focus on the coronavirus pandemic since it is a far more immediate threat to the public and world at large? Let's solve this crisis first. And then let's focus on other big problems before we worry about a long-term issue like climate change.

Absolutely and unequivocally on the first point. I pray that when you read this, the world will have already turned the corner and the global pandemic of COVID-19 will be under control. However, as I write this, that day seems too far off. We still need to win the war against this pandemic. We still need to shore up our hospitals and healthcare systems for a possible second and third wave of mass infections as the coronavirus makes its rounds to the southern hemisphere and back again in late 2020 and possibly early 2021. We still need to become more disciplined about social distancing, wearing masks, and washing/sanitizing protocols even after this virus appears under control. And we still need a vaccine as soon as possible.

But the reality is that climate change is an even greater existential threat to humankind than this coronavirus. The only difference is that it is happening in slow motion. We therefore need to lead on this emerging crisis in parallel with other immediate threats. We cannot continue taking the serial approach and allow climate change to be the honorable mention of strategic priorities every year. We can and must walk and chew gum at the same time. In fact, the global economic, security, health, and environmental consequences of not doing this are already starting to mount with thousands of lives lost and billions of dollars of property damage every year. And they will only get worse in coming years and decades.

That is why I am advocating for conservative leaders across the United States to renew their commitment to our future and common good. It is when we work together on our most difficult challenges and threats—even with people we disagree with—that we achieve our greatest contributions to mankind. America is great not because of any individual presidency or political party but because of our collective values, ingenuity, and courage to do the right thing.

Five years ago I wrote a book for CEOs to educate and inspire them on the strategic opportunities of clean energy in the twenty-first century. The premise was—and still is—that nothing is more important for their brands than leading with integrity. And given the preponderance of science and evidence on the growing threat of climate change, how can CEOs and boards of directors not want their enterprises to get on the right side of history and be part of this solution?

CEO Power & Light: Transcendent Leadership for a Sustainable World was laser focused on the top 0.1 percent who can do the most good the most quickly. Businesses are far more efficient in marshaling resources and achieving results than bureaucratic governments. Indeed, since I wrote this book, countless CEOs have seen the light … and are being the light to their companies, industries, communities, and the world at large. They are truly transcendent leaders! And this gives me hope.

However …

Climate change is hurtling toward us faster than ever. The scientific community is now saying the point of no return is coming within years, not decades. If we don't take bold action now, planet Earth will only become exponentially warmer over time, and the long-term effects will be catastrophic for life as we know it. And scientists are saying this—not corporate lobbyists, lawyers, and politicians—with greater certainty than ever before.

What do we do? Continue to implore government to get its act together—in vain? Conveniently assume other countries or companies will solve this for us? Pretend we do not know enough of the facts to make a commitment and difference? Hope our kids and grandkids will forgive our ignorance and/or arrogance amid the bleaker future they will undoubtedly inherit?

Most conservatives believe in the power of innovation and free markets to improve our quality of life and offer the chance at prosperity. Most believe in the importance of fiscal responsibility and not leaving our children and grandchildren with trillions of dollars of debt. Most believe in a strong national defense and defending our way of life at any cost. And most believe in the sanctity of life, whether it be in the womb, the last stage of life, or the promise of future generations. These principles and values have helped raise America from a colonial experiment to the greatest superpower in human history.

But we must continue to learn and grow as a nation while holding these values and principles near and dear to our hearts. In fact, any individual, company, or organization that does not learn and grow is doomed to fail. In other words, success is a function of evolution and sometimes reinvention, while failure is a function of intransigence and often arrogance. These are lessons I have learned in both my personal life and professional career.

It's against this backdrop and perspective that I present to you my new book.

Fusion Capitalism: A Clean Energy Vision for Conservatives is an update to many of the topics covered in my previous book and a wider appeal to all Americans, particularly conservatives. It's also a personal story of my quest for success and how happiness became defined by purpose, not just profits. I hold the conviction that our American economic system is uniquely capable of solving the world's greatest problem: climate change. Despite its weaknesses, which I am quick to point out, capitalism has improved our quality of life over generations better than any other economic system ever tried. Greed may be an unfortunate offspring of this system, but so is employment, education, and philanthropy. There is a reason America is considered the land of opportunity … and in many ways the envy of the world.

Of course, you already know about capitalism and its power to drive national and global economies, as well as mainstream new technologies, at warp speed and efficiency. You already know about its pros and cons. But why *fusion capitalism*?

Our sun is a million times larger than Earth and radiates more energy onto Earth every second of every day than what we humans have consumed over the millennia of our existence. It has been burning for billions of years and will burn for billions more. Its source of energy is a thermonuclear reaction called *fusion*, in which hydrogen atoms collide with great speed and become fused, releasing enormous amounts of new energy in light and heat. This happens ninety-three million miles from us. Yet the energy released is so abundant and powerful that we need hats, sunglasses, and sunscreen to protect our bodies. What's more, it is free.

Why is this relevant to capitalism? Because energy more than any other input to our economy is what powers our homes, cities, and way of life. Energy is the lifeblood of our very existence. We draw upon a wide, complex web of energy sources to power our lives, and yet nothing is more predictable, risk-free, and cost-free than the sun rising every morning and giving us more energy every day than we'll ever use. Why mine, transport, burn, and dispose of limited fossil fuels found underground when the costs are higher than capturing unlimited and free renewable energy above ground? This is not to mention the escalating external costs to our security, health, and environment.

In fact, fusion is widely considered the ultimate energy source.

Several countries are investing billions of dollars to tame the power of "stars" in laboratories. But some experts say it could take ten to thirty years to develop these technologies and make them economically viable. Others even say the energy to start and sustain this type of reaction will always be more than the energy generated. Given the existential threat of climate change, we don't have the luxury of waiting decades to hope for a new and better solution. The sun can and will serve us perfectly.

In fact, fusion from the sun is our best option—not fossil fuels, not man-made fusion, and not man-made fission (more than enough evidence on the disproportionately higher costs and dangers of fission prove this is true). Our future is just clean, abundant, free, and forever sun power. Sun power will provide us many times the amount of energy the human population on Earth will ever need in raw heat and light, and many times the amount of electricity we will ever need.

What we need to fully harness the sun's power and retool our energy economy is the biggest technological and economic revolution in human history. And I believe this is exactly the kind of challenge America is built for!

The steam engine is widely viewed as having launched the Industrial Revolution of the 1800s. The power that steam engines unleashed provided a platform for many new industrial technologies in factories and farms such as water pumps, locomotives, and mining equipment, which freed production from the limitations of manpower.

9

The electrification of homes and cities is widely viewed as having begun the electrical revolution of the 1900s. Just as steam power had done on an industrial level, electricity powered new household technologies like light bulbs, radios, TVs, refrigerators, and washing machines that freed individuals, particularly women, from countless time-consuming and tedious domestic tasks.

We are now more than primed for the technological revolution of the 2000s. Of course, it is already here with the internet, machine learning, and artificial intelligence. But I believe energy will represent the greatest opportunity and will always be the platform upon which most new technologies depend at their core. And now that the world population has ballooned to seven billion, it's time fusion energy (from the sun) is harnessed to its full potential.

Don't just take it from me, though. Take it from the most successful business leaders in the world, from companies like Apple, Google, Facebook, Microsoft, Walmart, Amazon, and many others. In fact, the top brands in most categories have already started a clean energy movement. They are either installing their own solar and wind farms to power their operations or contracting with utilities and clean energy developers to purchase the solar and wind energy that will power their companies going forward.

Why? They want to build their brands into great ones. They want favorable PR and a mass following on social media. They want to hire the best and brightest talent. They want to erase the liabilities of fossil fuel investments from their balance sheets. They want to minimize the costs and risks of future lawsuits. But most of all, if they are authentic, they just want to do the right thing, because that is what great leaders do. We just need to follow that lead and then lead ourselves to help turn the clean energy movement into a clean energy revolution!

Though the United States barely cracks the top-ten list of clean energy economies in the world, and China is handing us our lunch in renewable energy exports, we can change this if conservatives embrace a fresh approach and rally around this once-in-a-century opportunity. The world population is going to increase another 50 percent within the next thirty years, and so time is of the essence.

This will not only be good for our national debt, deficit, and trade imbalance; it will be good for our global economy, security, health, and environment. Imagine the cumulative economic boom resulting from every company in America investing in and saving energy. Imagine the reduced threat of wars and terrorism that will result from reducing the demand for, and thereby the price of, oil in unstable parts of the world. Imagine the benefit to millions of people who suffer from asthma and other respiratory diseases of the elimination of coal-fired plants and oil-burning vehicles. And above all, imagine the countless lives and generations saved, not to mention the world's coastlines and infrastructure, by mitigating climate change such that feedback loops do not overtake and accelerate the warming process.

Fusion capitalism is not just my story; it is a story about all of us, about the human race, and the incredible planet we share. Some of what I have to say is sobering. I am not one to pull punches where science and facts disprove personal opinions and political biases. But I am also not one to hold back on recognizing and seizing strategic opportunities that are becoming increasingly obvious. The future that I see for America and indeed the human race is bright and abundant, but to reach it we must remove the scales from our eyes.

I was spurred to write this book out of a spiritual conviction and love for my children. I am also motivated by the strong belief that many of us who identify with conservative values have, up until now, missed the bus on clean energy. Conservatism, as I've always understood it, stands for, well, conserving … preserving our most important values as well as competing and innovating to have a chance at the American Dream. And I've found great happiness and success applying these tenets in my own life and business.

Clean energy is our future. And embracing this future is completely aligned with American principles, family values, and conservative ethics. It is completely aligned with the health and happiness of all Americans, including neighbors and family members whom we don't always agree with. In other words, this is one of the largest issues where the right and left can and should unite, with all the enthusiasm and commitment we can muster. Because it will be a war far more real than the current coronavirus pandemic if we don't. It will be a war far worse and longer than any we have fought on the battlefield. The laws of physics tell us this … and we are just starting to see the early signs of it.

Fusion capitalism. The greatest economic system ever developed by man combined with the greatest energy source in our solar system created by God. And the coming together of America and humankind with a shared vision. It's time for the clean energy revolution to begin!

As for my electric car and home solar panels in my quaint, traditional community—they're not a statement; they're a promise.

PART 1
MELINK STORY

CHAPTER 1

Chasing Success

I woke early as usual. *God.* I sighed. *It's only 4:30 a.m.?* I lay there exhausted, mind racing, trying to solve the Rubik's cube of problems awaiting me at the office. *Just one of these days,* I thought. *I wish I could sleep until the alarm actually goes off.*

But surely I wasn't the only entrepreneur suffering from a hyperactive mind and its inevitable toll: a chronic lack of sleep and deep fatigue. I tried to rise out of bed slowly, not wanting to wake Mary Frances.

"Honey, is everything OK?" she asked in her characteristic calm voice.

"Can't sleep, Babes." I lay back down to feel her soft hand on my shoulder. "I have a thousand things to do and question whether I should drive to Cleveland for that conference this afternoon."

Second-guessing myself and how I might best spend my time

was a daily, if not hourly, ritual. The sustainability conference I was planning to attend would be a relative luxury compared to the gauntlet of customer calls, manager meetings, hiring and firing decisions, and design and production challenges facing me at work.

Mary Frances didn't say anything but started rubbing my back. Her gentle touch reminded me I was not alone. I could feel the stress melting away.

Not wanting to be sidetracked from my focus, I powered out of bed and jumped into my routine: shaving, showering, dressing, and going downstairs to make some coffee. When I stepped outside to get the paper, I saw the beginnings of a perfect June morning. A crimson ribbon of sunlight was breaking along the eastern horizon to a chorus of robins. *Ah, this is the early riser's reward*, I thought.

Sitting down to coffee and peanut butter on toast, I scanned the front page. More bad news in Iraq. *Darn it, I feel bad for our troops and all the families suffering over there*, I thought. Roadside bombs were taking a daily toll on the hope for a new government and way of life in that crazy, fractious part of the world. "We have so much to be grateful for," I said aloud in the silence of my kitchen, as if to emphasize the point. "But we can't take it for granted."

I decided to go to Cleveland after all. I had been reading about the sustainability movement and the LEED rating system for too long to ignore this opportunity. LEED stands for Leadership in Energy and Environmental Design, and it was all the rage for those of us working in the building industry. It seemed these types of conferences in the past were mostly on the east and west coasts, but this one, being in the Midwest, was too close to pass up. Every trade publication was hyping the cause of designing a more sustainable world, one building at a time.

My company, Melink Corporation, provided HVAC testing and

balancing services for national chain accounts and manufactured an energy-saving product for commercial kitchen ventilation systems, and so I wondered what small role we might play in the green building space. Should we just continue to sell our products and services as solutions for a need, or was there a bigger and more strategic play? I didn't want to get left behind as the building industry reinvented itself for the twenty-first century.

By 6:30 a.m., I was on the road. Interstate 71 North is a straight shot from our home in Cincinnati to Cleveland passing through Columbus. It's a road I know well, so I could set the car and my mind on cruise control. My busy schedule left me little time to reflect on the past and plan for the future, and so I looked forward to the easy five-hour drive and some respite from my normal pell-mell pace. I took in the peaceful rural landscape, a hazy morning sun shining on mile after mile of fields planted with corn, soybeans, and hay grass. *God's country*, I thought.

I had left the house before the kids were up. Mary Frances, still in her robe, was making lunches for them. I knew it wasn't just problems at the office that had made me hesitant about my trip; it was also losing precious time with my family. Another cost paid for running and growing a business—kids who are used to Dad being gone by the time they wake up. But then again, had it really been so different for me as a boy? Not really. My father had paid the same price for his career and providing for us. Actually, both my parents taught us the value of hard work and the sacrifices and virtues that often sprang from this ethic. I could only hope I was doing the same for my children and that they would come to understand.

I was the middle child in a family of nine kids—four older and four younger. Under the circumstances, developing survival skills was not an option. My dad was a successful engineering manager at

General Electric, but that did not create material abundance for such a large clan. My mom stayed home taking care of important things like keeping the house organized, washing our clothes, and preparing our meals. Somehow, together, they managed to send every one of us to Catholic grade and high school, and some even to private colleges. For my parents, a good education and moral foundation were more important than new clothes and toys.

While Dad was the type A, hard-charging perfectionist, Mom was the loving, caring listener. They complemented each other very well. We all needed a healthy dose of both discipline and empathy. A big family, with only one shower and a single TV, meant lots of sharing, compromising, and, yes, complaining. It also meant lots of hand-me-downs, rarely going out to eat, no air-conditioning during the hot summers, and making do with one station wagon to haul us around. But we all knew we were incredibly lucky and blessed. We lived in a big white house in Loveland, Ohio, northeast of Cincinnati, with a big lawn shaded by hickory trees, two acres of woods in the back, a horse farm on the right, and a tree farm to the left. Despite the hyperactivity and noise of a large family, we grew up with incredible love and respect for one another.

I took piano lessons and played baseball, football, and basketball growing up. Fortunately, I also had a day job called school. I loved school. Studying, learning, and immersing myself in the quiet of my own thinking were strangely fun to me. To this day, I can remember every one of my teachers' names going straight back to kindergarten. Few experiences were as emotionally rewarding as getting As in math and science and showing Dad when he came home. These subjects were a ticket to a good future in his mind, whether as a doctor or engineer, and I wasn't going to let him down.

Soon, I was off to Archbishop Moeller High School. At Moeller, football was king, but being too small to play competitively there, I gravitated toward wrestling. My older brother Jack had wrestled four years ahead of me, so I got to learn lots of moves and holds from him. Plus, I enjoyed lifting weights and got pretty strong for a kid my size. By the time I was a freshman, I was wrestling junior varsity and eventually made it to the state tournament as a senior. More importantly, wrestling was one of those paths of most resistance that helped forge the person I was becoming. The conditioning and training were one thing, but losing weight from an already lean frame to wrestle at the lowest weight class possible was one of the hardest things I've ever done. I believe the self-discipline it took to say no to the calories my body screamed for every minute of the day during the season gave me a mental edge in other endeavors.

Still, academics remained my true passion, and so I was more competitive in the classroom than on the wrestling mat. I was not often the smartest in the class, but I knew I was one of the hardest working. Somehow, I found time to serve as junior class president and vice president of student government my senior year and still graduate fifth in my class.

In 1976 there was a great deal of excitement and celebration in the air. Our country marked its two-hundredth anniversary as the world's first democratic republic. Concorde began the first supersonic transcontinental service to Washington, DC. NASA unveiled *Enterprise*, the first space shuttle, just as its Viking I probe landed successfully on the surface of Mars. Two young entrepreneurs named Steve Jobs and Steve Wozniak formed a computer company under the quirky name "Apple." And our beloved Cincinnati Reds were on a historic tear, on their way to defeating the New York Yankees in the World Series.

It was an auspicious year in my personal life too. In March I received an NROTC Scholarship to Vanderbilt University in

Nashville, Tennessee. The scholarship would take a lot of financial pressures off me and my parents, I knew. And Vandy would allow me to explore the South, when my older siblings had either gone west or north for college. It would also let my dad live somewhat vicariously through me, because he had gone to Vanderbilt for a short time prior to entering the European theater in early 1945.

I always believed, to some degree, that the shortest path to success was the path of *most* resistance. For this reason, along with the fact I excelled in math and science and that my dad was an engineer, I decided to major in mechanical engineering. Little did I know it would be almost as hard as losing weight for wrestling. While many of my Sigma Nu fraternity brothers were at the house during evenings having a good time, I was almost always in my dorm studying. After my sophomore year, I decided the navy was not for me, and so, on top of a tough course load, I had to finance the second two years on my own through hard-labor summer jobs, loans, and serving as a resident advisor in the campus dorms. The road to success seemed to get progressively steeper, but I continued to love learning, kept climbing, and in the end graduated cum laude in 1980.

I was already thinking about getting my MBA, but I knew it would be important to first gain some real-world experience. I also needed to pay off my undergraduate college loans and save money for tuition and living expenses. So that spring, armed with my bachelor of science degree in mechanical engineering, I interviewed with over ten companies, including energy giants like Exxon, Shell, and Duke Power. But my heart was with a midwestern heating and air-conditioning manufacturer called Trane, based in La Crosse, Wisconsin.

At Trane I worked with an engineering team on developing the next generation of commercial heating and air-conditioning systems. The goal was to improve energy efficiency and reliability and compete more effectively with Carrier, our main industry rival. But the main takeaway from this experience was, though I got to work with some of the nicest people I have known, that I did not want to spend a career doing this type of work. Though I was the first to arrive in the office at 7:00 a.m. each day, I was, honestly, bored. I couldn't help but watch the clock until it was time to leave at 4:30 p.m. I also realized that my dreams of climbing the corporate ladder and eventually assuming the leadership of a company were not likely to be realized working as an engineer. I was impatient. I wanted my MBA as soon as possible. Management was the ticket to success, in my mind.

Within a year I was applying to top-tier schools like Harvard, Yale, and MIT, but I wasn't accepted. Of course, I was disappointed but accepted the fact that I needed more work experience. So I stayed a second year at Trane and committed myself to improving my GMAT score. This time, however, the South was beckoning to me after a couple harsh winters in the upper Midwest, and so I applied to Duke University, University of North Carolina, and other top schools in the Southeast. When I received a scholarship to Duke's Fuqua School of Business, I packed my bags, bid farewell to my friends at Trane, and left La Crosse for Durham, North Carolina.

My time at Duke was critical both professionally and personally. Professionally because I had the opportunity to meet and study with some of the smartest people I had met up to that time. We were all type As, focused on career success and determined to get that ultimate competitive weapon in the business world—an MBA degree from a top-tier school. Even more importantly, I think, this mind-stretching, stimulating experience gave me the confidence that I could go out and do almost anything in the business world.

But this was an even more defining period in my personal life. Between my first and second years at Duke, I got an internship with

Burlington Industries in Greensboro, North Carolina. One evening, I went to a bar recommended by my landlord. I sat at a corner table reading a business article for work over a burger and cold beer. When I finally laid down my pen and put the article away, I looked up for the waitress to get the check and saw that, while I had been absorbed in *Principles of Financial Reporting,* two girls had sat down a few tables away. The girl on my left, a natural beauty in a light-blue dress, with dark hair pulled back by a headband, startled me. It wasn't because I felt like I knew her but because I definitely felt like I had to. She turned her head and saw me looking at her, and I was sure, though she denies it's true, I saw a hint of a smile before she turned away.

I don't know how I found the courage, but I soon got up from the table and approached her to ask for a dance. She introduced herself as Mary Frances Marsicano. She had just graduated from East Carolina University with a nursing degree, and I found her altogether captivating. She was poised, laughed easily, and seemed to look right into me with her large brown eyes. After that first dance, I knew she was the one. We dated through my second year at Duke, and that spring I popped the question. I'm happy to report she said yes and that it was the smartest decision I ever made. To this day she is my best friend and soul mate. And in 2020 we celebrate our thirty-sixth wedding anniversary.

Upon graduation in the spring of 1984, I accepted a job offer from Emerson Electric in Saint Louis, Missouri. It seemed like a good fit because, as a component supplier for the HVAC industry, the company appreciated my experience with Trane, and I was grateful to be able to leverage my HVAC background rather than have to learn a new industry from scratch. As a program manager, I was charged with cost-reducing a line of expansion valves, which gave me good exposure to manufacturing, labor relations, and accounting.

But my experience at Emerson was not a positive one. I quickly became frustrated by the bureaucracy and unhappy with the culture. There was an obsession with planning and not doing—in my way of thinking—and there were too many people undermining rather than supporting one another. I remember working on the factory floor one Saturday morning, organizing a response to thousands of expansion valves that had failed various tests along the production process and getting cold stares from the union employees. I was good at organizing, and if I didn't take the bull by the horns, all the excess inventory would only grow into greater chaos with time and bog down our whole system. Sure enough, I was hit with a union complaint on Monday and told to back off and stop trying to do the line workers' work. I was trying to lead by example and show that as a manager I was not above rolling up my sleeves. But apparently upholding the divide between management and labor, and their mutual lack of trust, was more important.

Similarly, I remember management implementing time studies on our production processes and holding the factory assemblers and testers strictly accountable for meeting the "faster" standards. Certainly this made sense for efficiency and costing purposes, but at the same time, managers weren't holding themselves to the same high standards. Every day I was witness to middle managers frittering away their time roaming the offices, chatting with others about personal matters while enjoying their coffee, and generally being very unproductive. I bristled at the double standard. Why weren't they working like they expected those on the factory floor to work? On every level the culture seemed too dysfunctional to fix, at least in my time frame. For the first time, entrepreneurial feelings I never knew I had were being stirred.

I realized that if I was going to start my own company, I needed practical experience in a small firm where I could be exposed to all facets of running a business. I remembered my dad's advice: "Steve, work in the trenches, and learn the business from the inside out." A sheet metal manufacturer, Maysteel Corporation in Mayville, Wisconsin, seemed the ideal opportunity to do just that when they offered me a position as their director of engineering for a line of commercial kitchen ventilation systems. I accepted and moved to nearby West Bend, Wisconsin, in January of 1986. Why I chose the cold north again after my earlier experience can only be explained by my desperation at the time. Mary Frances followed me a couple months later, after selling our house and quitting her nursing job at a nice hospital in Saint Louis. Only in hindsight can I imagine that being married to me was not the secure and stable existence that she had undoubtedly hoped for.

Unfortunately, I again found myself in a situation where the lessons I was learning largely sprang from seeing how an operation should *not* be run. As a manufacturer, we were much more focused on shipping product out the door and making our numbers than making sure the product met our quality standards and was not going to cause future problems for our customers. And even more troubling was the fact that I found myself in another unhealthy company culture. The lack of respect that colleagues showed one another was appalling and ran counter to what I had grown up with and been taught most of my life. Outright shouting and cursing reverberated in the conference room next to me every Monday morning, as numbers from the previous week were unveiled and blame was assigned.

But my experience at Maysteel had an unexpectedly positive, life-changing outcome. I got the idea for the business I would soon start. I was constantly getting calls from customers complaining that their ventilation systems were not working properly. Their kitchens were too hot, too cold, or filled with smoke. It didn't take long to figure out that this was happening because they were not air balancing their HVAC systems. I saw an opportunity and chance I was willing to take.

This was a pivotal time in my life. Here I was, a career-minded professional with engineering and MBA degrees from top schools, with substantial work experience, and I was contemplating doing something highly risky and almost unbecoming for someone with my background and credentials. What if I failed? How would I ever rebound? Was the path of most resistance worth it, again? Would Mary Frances begin losing faith in my ability to provide for her and our future family?

I started Melink Corporation as an HVAC testing and balancing firm on February 15, 1987. Of course, I remember the day as if it were yesterday. Upon waking up and reminding myself I was effectively unemployed, I put on my jeans and flannel shirt and went downstairs. Mary Frances kissed me on her way out the door to the hospital where she worked in Milwaukee. I sat in the kitchen contemplating the fact I had nowhere to go and finally decided to attend the 8:00 a.m. mass at the church across the street from our apartment. I needed help from a higher source. After mass, I went back to my apartment, lay down on the couch in our family room, and broke down. I had just made the biggest mistake of my life, I was sure.

CHAPTER 2

Finding Purpose

It takes courage to grow up and become who you really are.
—E.E. Cummings

W hen I heard the low-fuel alarm ping, I was suddenly jarred from my reveries. I looked at my gas gauge and realized (again) that my Ford Explorer was drinking gas like there was no tomorrow and I needed to fill up. In the distance I saw an exit sign and could see the Exxon and BP station signs beyond and to the right. *Ah, this will give me an easy off-ramp and on-ramp back onto the highway,* I thought. *Plus I can stretch my legs a bit.*

Ten minutes later I was accelerating back onto the interstate. *Good thing my gas tank needed filling,* I mused, *because my bladder needed emptying even more.* The sun was arcing higher in the sky, and the morning dew had given way to a noticeable humidity in the air.

It was only midmorning, but I was already halfway to Cleveland. The drive was flying by, and I resisted the urge to call in to the office. They knew how to reach me if they needed to, and I was enjoying the luxury of letting my mind wander, replaying mostly fond memories of the road I'd traveled to get to this point in my life.

Thinking back to that first day of starting my business, I reminded myself that the key to success as an entrepreneur is resilience. After initially feeling sorry for myself for giving up my stable job and becoming a wildcatter, I went upstairs to the spare bedroom and started thinking … thinking about how to start this so-called business. Then I went into action and started doing … a gut instinct of every true entrepreneur on the planet. I picked up the phone and did something I had never done before. I started calling mechanical reps and restaurant chains to sell my air balancing service. Within weeks I had landed a few small jobs and slowly acquired the art and science of air balancing.

Fortunately, I had a vague notion of what was required from my experience at Maysteel but had never done anything like this myself, especially so hands on. Basically, I had to climb a ladder onto the roof of a restaurant and adjust the motor pulleys of exhaust and make-up air fans, and then go down inside the kitchen to measure the resulting airflows. The goal was to ensure the fans ducted to the kitchen hoods were moving the correct air volumes for proper smoke capture and optimal energy efficiency.

My focus was on restaurant chains since I knew they had kitchen hoods, and smoke capture and negative air pressure problems. And getting repeat business from chains sounded like a sustainable business model. In fact, I knew people from Wendy's and International Dairy Queen from my work at Maysteel, so these were logical places to start building my customer base.

However, West Bend, Wisconsin, almost an hour north of Milwaukee's airport, was not an ideal place from which to fly around the country doing this work, and so five months after I started my business, Mary Frances and I relocated once again—this time to Cincinnati. Of course, Cincinnati was my hometown, but it also got Mary Frances eight hours closer to her family in Greensboro. More importantly, it allowed me to take advantage of the growing Delta Air Lines hub, which offered nonstop flights to most medium and large cities across the country.

It goes without saying that taking the entrepreneurial leap meant going from earning a good salary to almost nothing for a second time in the last five years. Fortunately, Mary Frances was very employable as a nurse—although nurses don't often get paid what they're worth. I remember going grocery shopping with her one day and putting some cheese in the cart. "Put it back," she said. Apparently, we couldn't afford cheese anymore. Peanut butter and saltines would have to do for a while longer.

Though my service offering wasn't exactly high tech, there was a definite need for it. My advantage was a broad knowledge of everything from engineering to business, HVAC, controls, and most recently kitchen ventilation. My competitors were mostly mom-and-pop contractors without much advanced education and training. Some of them would test smoke capture at the kitchen hoods with cigarette smoke or toilet paper. My state-of-the-art equipment and meters, by comparison, impressed the construction and facilities managers I worked with, and I provided a new level of professionalism, accuracy, and consistency to their air balancing needs.

I started out as a one-man operation, working out of the basement of our new two-bedroom home in Cincinnati. One day I would fly to New York City to air balance a Wendy's and the next day fly to

Orlando, Florida, to air balance a Chili's. Then I would come home and work on the reports, send them out with the invoices, and hope and wait for a check in the mail.

In 1988 I became certified by the National Environmental Balancing Bureau and became a licensed professional engineer in the state of Ohio. These credentials allowed me to further differentiate myself from the average competitor. A lot of national restaurant chains were rapidly expanding during these economic go-go years, and so I was able to hitch my wagon to their horses. I was soon working with Golden Corral, Bob Evans, Outback Steakhouse, Darden Restaurants, and many more.

People would often ask me how I landed so many national accounts working from my basement. I would reply that there is an amazing thing called a telephone, and, by golly, I worked it like the devil. Cold-calling construction managers who were living the problems I was uniquely able to solve for them was a necessary survival skill. After sales of less than $100,000 the first year, I grew revenue to about $250,000 the next. Major milestones included hiring my first technician and buying a copier so I no longer had to go to the library and insert quarters and dimes to make copies of reports.

But most of what I made from my service was about to be invested in my next big idea.

The idea came to me in 1988 while air balancing a Wendy's on a hot summer day. I was measuring the rotational speed of an exhaust fan and noticed the air coming out of the fan was a nice and cool, comfortable temperature. In other words, there was no cooking taking place inside the kitchen—but the fan was running at full speed and "throwing away" a significant volume of conditioned air. This struck me as extremely energy inefficient. Why not automatically slow down the exhaust and make-up air fan speeds when there is no cooking taking place? This would save fan energy and, more importantly, the conditioned air that was being exhausted and wasted.

It was a lightning-bolt moment. My heart raced … and my brain followed. I was vaguely aware that variable-speed technology was becoming more prevalent in the HVAC industry, and this seemed to be another good application for it. No, a great one!

I applied for a US patent and got it. Then, I applied for a Department of Energy grant and was awarded $88,000 to commercialize this invention. This was a lot of money for someone working in his basement at the time. Even the federal government thought it was a great idea! So every spare moment I had between air balancing restaurants, I was planning, developing, and prototyping the first variable-speed controller for commercial kitchen ventilation systems.

I was soon spending thousands of dollars on literature and buying component parts for testing. Then I was spending tens of thousands on trade shows and production equipment. While half of my basement remained an office, the other half became a factory floor—with a drill press, punch press, and lots of inventory. How Mary Frances put up with this encroachment on our living space I will never know.

In fact, Mary Frances was wholly supportive of my test and balance business. Every job meant more revenue and cash in our pockets. But this invention thing was potentially a hole to nowhere. It was sucking real-life cash from our slow-growing standard of living at the same time we were starting to have kids. She did not want me to return to the old days of working for a corporate salary, but I knew she wasn't convinced that I should run after my dream of becoming a manufacturer, just when my service company was doing so well. But she somehow kept faith in me and patiently supported my latest ambition.

Given how much energy restaurants use, I was expecting the industry to totally embrace my new product. I remember walking away from my first trade show thinking I would soon sell thousands of these systems. But things happened much more slowly. The market was not quite ready, and in reality, my product was not either. Codes had to be changed, end users and engineers had to be educated, and hood manufacturers had to be partnered with. I learned that the status quo was my biggest competitor of all.

In fact, the harder I worked and the more I invested in getting this second business off the ground, the quicker and easier the first business grew. The restaurant industry knew it needed my air balancing service because it was often required by code; plus, the safety, energy, and comfort costs of not getting the service were far greater than my fees. And being certified, independent, and national allowed me to keep landing new accounts.

But I kept at the second business—for years. One installation at a time, I would learn how to improve the product. One month it would be at a restaurant kitchen, then a hospital, then a supermarket, then a school. Slowly but surely I managed to breathe life into this most stubborn business.

In the meantime, Mary Frances and I were growing a family. In 1990, Matt was born, then Monica in 1991, Katie in 1994, and Laura in 2001. Mary Frances switched from being a full-time nurse to a full-time mom, with occasional part-time work as a continuing education instructor at the local hospitals. Needless to say, the pressure was on me to become a dependable provider for my family.

Over time, my basement became too small, and I had to expand into my garage. In 1993 I picked up an old hood system from a White Castle restaurant and installed it in my garage for testing purposes— along with a large exhaust fan sticking out above the roof. We lived on a busy street, and to this day I am amazed my wife went along with it and my neighbors didn't complain.

The fact that I had accommodating neighbors was proven again later when the two businesses had grown enough to justify moving them into another house down the street. I needed a zoning variance to put an office there. If the city had known I'd be doing manufacturing in the basement, I don't suppose they would have allowed it.

I eventually realized that to take my variable-speed controller to the next level, it would have to be redesigned from the ground up. And that would take a major investment. I remember thinking it could go great, or I could go bankrupt. But I didn't want to make just an incremental improvement. I wanted this next-generation system to be so well designed that the hood manufacturers would want to start building it into their ventilation systems for new construction. Up until then all our installations had been retrofits.

This was a strategic decision. Should I continue to invest my life's earnings in this second business, which was growing much slower than I would have liked? Or should I cut my losses and focus on my testing and balancing service business, which was very profitable?

It came down to answering some soul-searching questions: Who am I? Why did God put me in this position? Is money the only objective here? What about making a real difference in my industry? Would I be content throwing in the towel when my idea had so much promise, even though that would have been the easiest thing to do? Could I be happy developing a product that might never be successful? What does self-actualization really mean to me?

After many long walks at night and looking to the stars for some divine truth, I found peace of mind. The mental convulsions were finally over. I realized that purpose is more important than profit and decided to double down and show the industry what the future looked like. I no longer cared as much about making money off this invention as I did about being true to myself and trying to do something meaningful and important. And perhaps I no longer cared as much about what others thought of me, realizing now that the opinions of family and friends I had often been seeking along the way should no longer be my North Star.

Mary Frances would surely balk, I thought, but she would hopefully ultimately support me. She always had. Making me happy was what made her most happy, I wanted to believe. God, I am lucky.

> I no longer cared as much about making money off this invention as I did about being true to myself and trying to do something meaningful and important.

And though my parents and brothers and sisters would surely wonder, "Why does Steve continue taking risks in his life that do not seem to pay off?" the safe and sure path was not mine to choose. In other words, I believed God wanted me to engage with life, face my fears, and fulfill my potential.

So I hired more staff. One was a graduating engineer out of the University of Cincinnati; Darren Witter is still with me almost twenty-five years later. Another was a salesman who called on me at the time; Ted Owen is currently one of our top distributors. Over time Bryan Miller, Jason Brown, and many others became great additions to our team.

I soon leased a manufacturing plant for higher-volume production. If we were going to gain the confidence and trust of major companies, we had to look and act the part. Working in the basement of a house down the street was no longer going to cut it.

We also gave our next-generation system a new name, Intelli-Hood, and came up with a new strategy. We decided to stop *pushing* our product to the hood manufacturers because they were mostly sheet metal manufacturers who did not understand and appreciate our controls. Instead, we decided to create market pull by going around them and selling directly to the end users and engineers.

We finally started getting traction. Supermarket chains across the United States and the United Kingdom started specifying our Intelli-Hood on new stores and retrofitting them on existing ones. Restaurants then started to do the same, then schools and colleges/universities.

Even codes became more relaxed. Over the years, to a large degree because of our perseverance, code language went from suggesting our controls were *not* allowed because of a minimum duct air velocity requirement to saying our controls *were* allowed because of the type of sensors we utilized, then to actually requiring our controls or similar energy-saving devices in states and jurisdictions that had adopted the most updated and energy-efficient standards. My gamble looked like it might finally pay off.

All along, our testing and balancing business had been growing steadily. We were hiring technicians in key markets and building a bona fide national network. This allowed us to reduce our travel costs when servicing stores across the country, and it gave us another leg up on our competition.

By 2004 we were operating out of two small buildings and a manufacturing plant. We were a medium-size business with approximately fifty employees, and all indications suggested we had a bright future ahead of us.

But …

In hindsight, I was approaching business like most CEOs. I was mainly in it to make a comfortable living and hopefully a fortune someday. But there was little crossover between our products and company culture and my personal life. Our products and services were designed to help save energy for our customers, but I was driving an SUV, and there was nothing about my home or offices that said I was committed to energy efficiency. The growing problem of climate change was not really on my radar screen, nor was the concept of sustainability or even green buildings.

I had gradually developed the managerial skills needed to run my company, but I had not yet developed the leadership skills needed to try to change our industry. I had never imagined the possibility that my life's calling was still to be found.

* * *

There were two distinct events in my life between 2002 and 2004 that profoundly altered my perspective, giving me a sense of calling I had now found myself searching for.

Dad had died suddenly in 2002. Seemingly in good health, he was working in the yard when he suffered a stroke. I had a world of respect for my father—not just for the man I had known as Dad but for all that he had overcome as a child and young adult and for what he became for many people who knew him. Dad's father was ill most of his childhood, so young Johnnie had to help his immigrant mother and four younger siblings adapt to life in the bitter cold of Hibbing, Minnesota, and survive the Great Depression.

Then World War II came along, and Dad volunteered to serve—a decision he confided he had come to regret, only because of the hard times it imposed on his mother, father, and sisters. After the war Dad came stateside again and earned his mechanical engineering degree from the University of Minnesota with help from the GI Bill. Then he started a career with GE. A blind date later, Dad met and soon married my mother, Jeanne. Being a devout Catholic, he and Mom started their large family.

Dad was a great mentor in many ways, and the abiding lesson I learned from him was to always strive for excellence. Though he'd been an easy man to love, at times he was a hard man to like, so I guess I'd been surprised at the profound effect his passing had on me. Not having the chance to say a final thank-you and goodbye was hard. But in the weeks and months afterward, I felt like a proverbial torch had been passed. There was a void of sorts, and I felt the almost physical urge to try to fill it somehow. Could I ever measure up to be the man my dad was?

I took a lot of satisfaction at that time in the fact that my company was thriving and I had a wonderful family. But I also felt that I was at an intersection, that I had come to a place where there was something else I needed to find. And then, the summer after my dad's passing, a business coach and friend suggested I read the book *Visioneering* by a minister and author named Andy Stanley. I can't say I was looking for a book on spirituality at the time, but I was open to something new, so I indulged him.

Stanley's book is about the unique opportunity leaders have to make a difference in people's lives. He makes clear that the power to create a vision for yourself and your organization is not to be taken lightly. Leadership comes with a megaphone to speak the truth as you know it. A leader who can lay out a great vision is not just a boss but a source of inspiration.

I remember reading the book in bed, with Mary Frances next to me, telling her about its effect on me. I could barely take in a page at a time without having to pause and absorb the challenge I felt being personally directed at me. Though I was a CEO in title, my company was small enough that I could still act more like a manager than a leader. I could strategize and plan but not often lead and inspire. As a history buff, I loved reading autobiographies of great leaders who exemplified wisdom and courage. Stanley was now challenging me to be one of them.

Visioneering was the affirmation I needed to become a better leader. It convinced me that God had given me a unique platform from which to influence others and make a difference. And it gave me the courage to not only think bigger but help others think bigger too.

I felt primed to lead something. But lead what? My company? How and why? Since when does a company inspire people to achieve something truly great? Companies are supposed to maximize profits … and what's so inspiring about that?

Traffic picked up as I approached Cleveland, and my mind turned back to the task at hand. I had a meeting scheduled with a client in the early afternoon, and then I wanted to check in at the conference, look over the schedule, and map out how I could best utilize my time. So often the case when I attended conferences, our team was setting up a sales booth on the exhibit floor, and my schedule was packed with back-to-back meetings. But this time I had the luxury of flying under the radar a bit, with time to pick and choose how to spend my time and attend sessions that interested me—to listen and learn.

And as it turned out, making the time to attend this conference made an enormous difference in my perspective and my life. I had been to many trade shows before, but those were mainly for the HVAC and food-service industries, and they were always fairly staid. This conference was different. The people I met—from architects and engineers to manufacturers and contractors—were literally intent on changing the building industry and making it more sustainable. For me, this was mind blowing. Changing the building industry and, in effect, the world? Now that was inspiring.

In my heart I knew the process of designing and constructing buildings was broken—too much focus on first costs and not enough on long-term operating costs or safety, health, comfort, and making the very buildings where we live, work, and play more inspirational for the human spirit. I was tired of being constrained by short-term and lowest-first-cost thinking, and what I was hearing in every presentation, meeting, and conversation at this conference was that I was not alone.

One presentation in particular stuck with me. It was by Kevin Hydes, who later became chairman of the US Green Building Council. In the course of his presentation, Kevin showed us a lot of figures and case studies, but a decade later, I don't remember any of those. What I do remember is his last slide—a photograph of him walking his daughter on the beach. She appeared to be about four years old.

Kevin said, "What it all comes down to is that I don't want to have to tell my daughter that I used the last barrel of oil."

Kevin's message and the integrity of his thinking resonated with me so powerfully, and I thought, *What a compelling message that last slide has for any parent.* I have four children—Matt, Monica, Katie, and Laura. At the time, the oldest was in high school, and the youngest was about the age of Kevin's daughter. Everything he had said about the calling we had as business leaders, Americans, and world citizens to usher in a new era of sustainability and clean energy had made sense to me. But the emotional chord he had struck with that final slide and his words of legacy stayed with me. I didn't want to have that conversation with my kids either.

I learned a highly valuable lesson that day from Kevin—the importance of storytelling within the context of our scientifically advancing world. It also drove home, on a visceral level, what it meant to live and work sustainably. More importantly, it made clear what it meant if we didn't.

This wasn't about cliché environmentalist talking points, nor was it a politically partisan message. It was about the lives and livelihoods of our children. What we did today would affect them for decades.

The question was whether our actions would leave them a world that is safer, healthier, and more prosperous, or poorer and less secure.

The conference gave me a lot to think about; it had been like *Visioneering* come to life. I did not know then that it would change my life, but it came at a perfect time. It was an illustration of exactly what the book talked about—the power of a common vision to inspire people to do their best work.

For me, it was revolutionary, and the next step quickly became clear to me. We were operating out of three separate locations and needed a new, larger building to consolidate and become more efficient. This was our chance to show the kind of leadership that Andy Stanley and Kevin Hydes were talking about.

"I am going to design and construct a supergreen building for my growing company!" I declared to myself at the end of the conference. And when I later learned there were only about one hundred LEED Gold–certified buildings in the world at the time, I was bound and determined that Melink Corporation's new headquarters would join that list.

Looking back, I realize this was my first inkling that I'd found a purpose bigger than myself and my company.

CHAPTER 3

Nearer, My God, to Thee

After the building conference, I raced home and shared my freshly inspired thinking with Mary Frances. Greeted by a warm hug and sweet kiss, I got right down to business. "Babes, you know how I've been talking about building a new building to consolidate our separate offices and plant?" It was a rhetorical question because I jumped in before she could answer. "Well, I want it to be one of the greenest buildings in the country. Do you know there are no LEED Gold buildings in Ohio yet? This is an opportunity for Melink to have the first one and to help lead the green building movement rather than just follow it!"

"Steve," she started. "Wow … I can tell something happened, and you're sold on this, but hold on. Are you thinking with your heart or your head?"

Still excited, I responded, "Both, I guess, but what's wrong with that? This world could use a little heart. In fact, the problem with the building industry is that everyone has been using their left brain to a fault. It's only about payback. That's no vision. Where's the inspired thinking about doing the right thing?" I explained why I believed my idea was not just the right thing to do ethically but also a strategic opportunity for our company to truly lead for the first time.

I could tell my passion alone wasn't going to sell her on my idea, and it ended up taking several conversations to finally get her support for making such a big bet on "going green." Even then, I felt like her support was tepid, clearly contingent on seeing results along the way. As it turned out, this was just the first of several encounters with critical stakeholders where I was going to experience pushback, some of it extreme.

The next step was selling my management team. After returning to the office on Monday morning, I asked my senior managers to clear their schedules that afternoon so I could share my epiphany and promote my new vision for the company. Having thought about it all weekend, I felt I was mentally and emotionally ready for the challenge.

By 3:00 p.m., everyone was in the conference room, some casually chatting and others simply waiting for me to get started. Normally, I would take my seat at the head of the table and begin the meeting with small talk about the weather or some factoid about our business or industry to break the ice and settle the room. Today, I stood.

"Team, there's only one agenda item today, but it's an important one. Over the past year or so, I have felt the need for greater purpose in my work … in our work. But it was not clear to me how that could best be manifested. Until now. Over the weekend I had a lot of time to think while driving to and from the LEED conference in

Cleveland, where I found real inspiration. I want to tell you about that and explain a vision for Melink Corporation that I hope you'll be as excited about as I am.

"Our company is in the business of selling energy efficiency," I began, "but I learned this weekend that we are behind the curve. There is a green building movement out there intent on changing the world, and we need to get on board by walking the talk on energy efficiency. The best way for us to do that is to lead by example … and so we are going to build the first LEED Gold building in Ohio for our headquarters, starting later this year."

I could tell my staff was a little shocked by the boldness of my announcement, and several faces around the room looked confused. So when I opened up the meeting for questions and comments, they didn't hold back. How much would this cost? When would we see a payback? Who was going to design the new building and decide what features to incorporate?

I saw a future where purpose and success could merge, but fellow managers who were closer to everyday needs and problems mostly saw the present. The resulting tension was real. Sure, a few managers seemed to embrace the new vision I was casting, or at least give it a chance. But the ones who were resisting were more vocal and would potentially be more influential with others in the company if I did not win them over.

Our operations manager, for example, said, "Steve, it's nice to save the world, be good citizens, and all, but I can't help but wonder if this LEED thing is just the latest environmentalist buzzword. I mean, we're a business. Will it put us at a disadvantage by sucking resources away from other priorities, like our employees?" I saw two other managers nod in agreement. I could extrapolate their thinking: *Yeah, sounds like it's going to come out of our bonuses.*

I tried not to get defensive, to let everyone have their say, and then tabled the discussion for future follow-up. *OK*, I thought as I walked back to my office, *consensus, let alone enthusiasm, is not going to come easy*. I had to remind myself that I had gotten to this point over time and after many life lessons, reflections, and a steadily growing trust in my faith. The building conference had just been the spark that had finally lit the fire. Was it realistic after one meeting with my staff to expect everyone to immediately feel the same sense of purpose and passion that I did?

So over the next several months, I invested the time to more fully educate and hopefully inspire my staff. I sent a number of managers to various building conferences so they could see for themselves how the industry was shifting inexorably toward sustainability. I wanted their buy-in to be genuine, not just a dutiful response to my marching orders. I would be patient, but I was also determined to make sustainability the new center of everything we did at Melink Corporation and to lead by example. There was a limit to how much I was willing to push uphill against those who just couldn't, or wouldn't, get on board. I ultimately needed individuals to lead, follow, or get out of the way. If maximizing sales and profits for their own benefit was the only mantra they could rally around, then it was time for them to move on and find another job. Our company now had a mission that was bigger than ourselves. It so happens the manager who pushed back at that first meeting never stopped and wasn't with us a year later.

> **Hiring people who shared the same values and vision would get us people who loved their work and bring an authentic commitment and loyalty to our newfound purpose.**

I learned a hard but valuable leadership lesson. We could no longer just hire people to fill a position; we had to hire people who believed in what we were doing and why we were doing it. Hiring people just because they're qualified to do a job would only get us people who want a paycheck. But hiring people who share the same values and vision would get us people who love their work and bring an authentic commitment and loyalty to our newfound purpose.

* * *

I began searching for an architect and builder who could help us achieve LEED Gold certification as soon as possible. At the time, that wasn't easy. Green building was still a new concept in the industry, particularly in our part of the country, and we quickly adjusted our requirements from "experienced in green practices" to "willing to climb the learning curve with us." After selecting a core building team, we met with other potential partners, from engineers and manufacturer's reps to different subcontractors. Integrated design was the name of the game. One idea built upon another. For example, a discussion on building orientation led to talk about natural light and insulation. Those considerations, in turn, affected lighting and HVAC design, and so on.

Overall, the process was methodical, collaborative, and even exhilarating. But there were some interesting headwinds too.

Our architect, for example, spoke excitedly about taking on a green project like ours during the bidding process, and after researching various LEED best practices delivered a solid preliminary design. But he did not appreciate that I, as the owner and customer, was also researching these best practices and had plenty of ideas on how to make his initial design even better. After all, I was footing the bill ... and the inspiration behind this project. His ego got in the way, and he chose to disengage from the project. It was too late to start with a new architect, and so I worked with the GC to implement as many of the improvements as we could. Ironically, that architect was quick to claim credit for the project when our building became a regional news story, even though he'd resisted the cutting-edge energy-efficient features that were creating all the buzz before jumping ship. A little annoying, to be sure, but also gratifying to see him and others finally coming around.

Another stark example of the almost knee-jerk resistance we faced came at the end of the project, during the plumbing inspection. The inspector looked at our waterless urinals in the men's restrooms and asked, "What are these?" I explained that they had been special ordered at considerable expense to help us save water at our supergreen building. I added that this relatively new technology was being used in thousands of locations around the country, including quite a few in Ohio. He matter-of-factly declared, "Well, not in my jurisdiction." In his mind, since he had never heard of them before, they were unproven and unsanitary, and he was not about to be challenged on what was allowable and not allowable as the authority on such matters. Therefore, he ordered us to switch them out for "normal" urinals.

We tried to appeal the decision locally but were told we would have to go to Columbus and appeal it at the state level. A couple weeks later I drove to the state capital for the appointed meeting, not sure what to expect. When I arrived, I saw the local plumbing inspector there waiting … he had also made the long drive and was going to fight me every step of the way. Fortunately, the board was very knowledgeable about the growing importance and popularity of green building and was sympathetic to me as a business owner trying to be innovative and invest in new approaches. Nothing ventured, nothing gained, was their opinion. I won the appeal.

Despite my victory, the plumbing inspector was determined to have the last say. Before he would sign off on our building, he insisted that we install a supply water pipe above each urinal. He was sure that, at some point, our illegitimate, abnormal waterless urinals would fail, and we would have to install "real urinals." To this day we have capped water lines sticking out of the wall of our restrooms. I've come to see them as badges of honor, monuments to innovation. Change comes hard for some people.

* * *

We completed the first LEED Gold–certified building in Ohio in late 2005. Our new thirty-thousand-square-foot headquarters cost $2.75 million, and the "green" features accounted for about 10 percent of that figure. Here are some of the facility's key "green" features:

- **A Superinsulated Building Envelope**
  If anything, I wish we had added more insulation. It has no moving parts and lasts as long as the building. It works twenty-four hours a day, in summer and winter, and requires no maintenance. It is the most important thing you can do to cut your heating and cooling bills.

- **Sunlight-Optimized Positioning**
  Assuring that our main facade has a southern orientation allows us to capture as much sun as possible. It provides natural heat and light and a much more pleasant working environment than a maze of fluorescent bulbs.

- **Geothermal Heat Pumps for Heating and Cooling**
  I wish they weren't underground so we could show them off, but the earth's ability to provide heat in winter and absorb it in the summer cuts our energy use by at least 30 percent.

- **Automated Fluorescent Lighting**

  Our lights—all fluorescent at the time—use less than one-third of the energy of incandescent bulbs. They operate on sensors that detect movement and natural light in a room. I love seeing the lights turn off automatically when someone leaves his or her office.

- **Automatic Building Controls**

  Automatic controls mean we're not heating and cooling to room temperature at times when no one is in the building. In the summer, for example, we set the thermostat to eighty-five degrees at night and on weekends then bring the temperature back down in time for employees to arrive in the morning. Similarly, our ventilation system is linked to a carbon dioxide sensor, so the amount of outside air we supply to the building only increases as necessary. The less outside air that comes in, the less we need to heat or cool it.

- **Solar Power**

  A solar PV array on our roof allows us to generate some of our own electricity on-site and reduce the amount we need from our local utility—another monthly cost savings that keeps on giving.

* * *

There were other smaller green features too. And while no single feature could be called revolutionary by itself, the sum total of these investments in energy efficiency and renewable energy made our building one of the greenest in the country.

I was proud we had achieved our goal of constructing the first

LEED Gold–certified office building in Ohio. Even more gratifying to see was that our team, despite the original pushback, was proud of the new facility too. Fittingly, Kevin Hydes, who had motivated me to take action with his presentation in Cleveland, flew in and served as the keynote speaker at our grand opening in early 2006. He seemed genuinely moved to have played a part in inspiring me as an emerging leader in the green building movement.

Kevin was joined by Ohio governor Bob Taft, state and local government officials, and professionals across the region and beyond—including Canada and the United Kingdom. I was humbled to speak in front of these movers and shakers from so many sectors of government and business, not only about our building but also about the need for reinventing the building industry. I did not realize at the time how often I would be relaying this message over the years to come, or how much bigger my goals would become.

I assumed that, after the hoopla of our grand opening, things would die down and we would go back to business as usual—more people efficiently working in one building and more energy efficient because of our green features, but otherwise back to normal. As it turned out, though, the new building kept us from getting back to normal in the most wonderful way. Maybe a better way to put it is that we reached a new normal. The publicity we received for having the first LEED Gold–certified office building in Ohio piqued people's interest, and a snowball began to roll downhill.

Our new building became a destination for anyone interested in learning the costs and benefits of going green. Primary schools, business executives, universities, elected officials ... just about every day, someone was calling and asking for a tour of our headquarters.

Though scheduling and giving these tours was time consuming, it was also extremely gratifying. Invariably, we highlighted the major green features of our building—the superinsulated envelope, the natural daylighting and fresh air strategies, and the geothermal heat pumps. But what most captured people's attention was the small solar PV array on our roof. Everybody from small children to politicians was fascinated by the idea that something as ordinary as sunlight could power our air-conditioning, lights, computers, and everything else that makes modern life modern. I don't blame them. I'm fascinated by it too.

All these tours inspired me as much as they inspired our visitors. Never before had an HVAC testing and balancing service and kitchen hood controls company become a popular attraction. Over time we have shown our building to thousands of people from businesses, schools, nonprofits, and government agencies who wanted to see what we had done and why. They wanted to learn lessons they could apply at their own places of work. And they were not all about energy efficiency and technology; they were also about values and culture.

Within six months of our grand opening, we knew we were on to something. All the attention we were getting was a sign we were doing something truly exciting and that we should continue pushing the envelope. In 2006 I learned about the 2030 Challenge proposed by architect Edward Mazria. His vision was that all new buildings should be designed to be net zero energy (NZE) by the year 2030. NZE means that a building generates enough of its own clean energy over the course of the year to offset any fossil fuel–based electricity or natural gas it might use.

This was bold and exactly the kind of challenge our newly energized company was eager to take on. This time when I met with senior staff to discuss the opportunity, I didn't have to evangelize. We all agreed we would set the goal to make our headquarters NZE by 2010—twenty years ahead of Mazria's timetable.

Pushing the envelope meant three things. First, it meant making a darn good building even better. For all the energy-saving features we had integrated into our new construction, we still had room to improve. Second, it meant showing how a more sustainable business translates into a more profitable business. Third, it meant continuing to spread the word about our newfound mission of sustainability and why we were committed to it.

We had already plucked the low-hanging fruit of energy efficiency in the original design, so getting better meant reaching for the mid-hanging fruit. By this I mean opportunities that have longer paybacks of between five and ten years. Of course, I was more confident in making these investments because I was seeing the payback on our initial investments every month. I was willing to accept a moderate 7 to 12 percent return on investment—especially during the Great Recession, when many other investments were yielding far less. Besides, I was planning to be in business at least another twenty years—and this was a tangible test of my commitment to putting our values first.

So we made incremental improvements wherever we could. We installed skylights to bring more natural light into the second floor and further reduce the need for electric lighting. We upgraded to LED lighting over time. We closed off the lobby at our main entrance so less outside air would infiltrate the office area. We installed interior storm windows to reduce heat losses during winter. We sealed the joint seams between the roof and walls to make our building as airtight as possible. We put in valves and controls to bypass our geothermal loop during mild conditions and so reduce pumping energy. We installed exterior light shelves over the south-facing windows to block unwanted heat gain during the summer, when the sun was high, but allow passive solar heating and lighting during the winter, when the sun was low in the sky. We added a battery storage system to reduce peak-demand charges on our utility bill. We purchased programmable outlets to turn off plug loads at night and during weekends.

We even bought insulated coffeepots so we weren't using energy to keep our coffee warm through the day. *If only we could convert the human energy expended on the treadmills in our fitness room into electric energy*, I thought.

On the renewable-energy side, we added more solar PV panels to our roof, parking lot, and grounds as we could afford them. They even generate power on cloudy days and look pretty cool too. We installed a special wind turbine to take advantage of the low wind conditions we have in the southwest Ohio River valley. It's the most iconic feature on our campus, and I love to watch the white blades turn slowly over the green lawn and fields behind them. And we put in a solar thermal system to generate free hot water for the showers in our fitness center. It's installed on a front-facing wall of our building, its shiny glass surfaces a testament to the fact that renewable energy can be architecturally beautiful and should not necessarily be hidden on the roof as a piece of industrial equipment. This is our mission, after all, and who we are.

And we installed two wood-pellet stoves—one in our Green Learning Center and one in our library—so we could augment our heat pumps using a renewable resource and create a warm glow in the office on cold, cloudy days. Our employees love them.

Some of this may seem slightly obsessive, but the benefits of working in a green building quickly became readily apparent. Not only were we saving energy, we were creating a healthier and more comfortable workplace ... and a happier workforce. Yes, believe it or not, endorphins are released when sitting next to a slightly open operable window where one can feel a wisp of fresh air and hear birdsong in the background. A comfortable and natural setting is healthy and promotes positive energy and productivity. I have yet to hear a plea for a return to the days of hot, stuffy cubicles and oppressive florescent lighting. In fact, many employees have told me they were often sick at their former places of work. A lack of fresh air led to high concentrations of carbon dioxide and headaches, dizziness, and fatigue. Though I wasn't initially planning on fewer sick days and more productive employees, this was another HR benefit I was happy to bank as our new normal.

After we implemented these improvements, we realized our LEED Gold certification was underselling us, so we applied for and received LEED Platinum certification under the existing-building category in 2010.

By the time 2011 rolled around, we were saving enough energy and generating enough of our own to claim our headquarters an NZE facility. This was big stuff, because in our search we could find no record of an existing building in the United States or even the world that had been retrofitted and improved to this high level of performance. To celebrate, we hosted an open house so family, friends, and professionals across the region could visit to learn and hopefully be inspired. We had beaten Mazria's goal of achieving NZE—a year late according to our own schedule but ahead of his schedule by nineteen years.

It is worth noting that getting to NZE didn't require any special expertise. It mainly took commitment and smart investments—about $750,000, or twenty-five dollars per square foot in our case. In fact, this amount would be much lower today, as many of our upgrades are now considered standard and the cost of solar power continues to drop.

If we were able to do this during the Great Recession, many more companies could do it now. I estimate the savings at $75,000 per year, so the simple payback period is around ten years. But in reality, with gas and electric rates likely to go up in that time frame and the fact that we financed the improvements, the actual payback period is probably closer to seven or eight years. And that's before accounting for the newly discovered HR and PR benefits, which, though not as easy to quantify, have undoubtedly added far more value to our company than the energy savings.

* * *

Which leads me to what has been the greatest impact and benefit of our company's sustainability journey—the effect on our employees, on the people who drive our company's success. Because they have been eyewitnesses to the interest in, recognition of, and validation of the fact that we are doing something good and important, they are energized about what they do. Not surprisingly, it has also allowed us to attract better talent than would have been possible in the past. Good people want to associate themselves with healthy, mission-focused companies. And because salaries and wages are the largest line-item expense for most businesses, a small improvement in employee productivity can dwarf the awesome energy savings we are earning every day, month, and year. These are the HR and PR benefits of building a sustainable, future-driven company culture. Again, they may be difficult to measure, but they are as real as money.

As our reputation for sustainability spread, we started attracting more and better applicants for every job opening. For example, we found Craig Davis, our current president; Randy Miles, our VP of marketing and sales; and Seth Parker, our VP/GM of solar and geo. Or rather, I should say these talented leaders found us. The same goes for Angela Bradley, our director of HR; Mike Murphy, our director of operations; Allison Sternad, our director of marketing; and Joel Geiman, our GM of Melink T&B. We also brought on Brian Ross as our CFO; Janice Scheid as our controller ... the list goes on and on. Not to mention all the key employees we have retained over fifteen-plus years, like Darren Witter, VP of HR; Bryan Miller, VP of technology; and Jason Brown, everyone's favorite utility player. In fact, another book could be written just about the people we have at Melink Corporation. Everyone, from our field technicians to the back-office staff, is top notch, and most of them would not be at Melink were it not for our culture and brand. Many of them could probably get paid more somewhere else, but there is something to be said for a workplace that is respectful and fair, healthy and comfortable, and committed to an important mission.

Of course, our enlightened business philosophy needs to be checked from time to time. So we conduct an annual independent company survey to see how we are doing. This feedback allows us to focus on areas needing improvement. As a result, we try to make what we consider to be the most important changes, and this whole process earns us respect up and down the chain of command.

Another principle we follow is that ego isn't rewarded here—pride needs to take a back seat to performance. Titles are not a privilege; they are a responsibility. Everyone has a say. As a result, backbiting and politics are rare, and if they surface, our team addresses them immediately. Respect is the watchword at all times.

As it turns out, some of those early experiences I had in dysfunctional companies turned out to be invaluable. It's great to see how things should be done, but it can also be valuable to have a front-row seat to how things shouldn't be done. A healthy, supportive company culture allows employees to focus on bigger things than internal drama and politics. As Brian Ross, our CFO, has proffered, "If we're not on a mission—which we are—we as humans have a tendency to fill the space with other pursuits."

One of the most important things I want to share with you in this book is that doing the right thing actually boosted our profits and enterprise value. Committing to a responsible, sustainable, mission-based business didn't just make us feel great; it coincided with a period of unprecedented growth in our company's history. As it turns out, there wasn't a financial price to pay for our higher purpose and values. The fact is, we would have paid a heavy price for continuing to operate without them.

First example, who wants to buy a building that's an energy hog? Studies show that high-performance buildings rent for 15 percent more and sell for 20 percent more than they would otherwise. The EPA estimates that every dollar spent on efficiency increases asset value by two to three dollars. That monetary value will only increase as electricity prices go up, as they are likely to do. The US Energy Information Administration projects them to rise by more than half a percentage point per year, after inflation.

Beyond the project's return on investment, our new high-profile reputation as a green building innovator elevated our brand. Because our existing and prospective customers have been eyewitnesses to the local, state, and national attention and awards we've received, they see us in a new light. We are more credible to them, and they are more interested—sometimes even eager—to do business with us. We couldn't have bought that kind of advertising before. Now, we are getting all this publicity for free. The result has been more and better networking, qualified sales leads, strategic partnership opportunities—and increased sales! The PR benefits are undeniably one of the major forces behind our rapid growth.

Being a national leader in sustainability also enables us to attract a whole new breed of customer. Today, we are working with some of the largest and most successful companies in the world, including Apple,

Google, Facebook, Walmart, Target, Whole Foods, McDonald's, Yum! Brands, Starbucks, Hilton, Hyatt, Marriott, Publix, Disney, Procter & Gamble, the United States Army, and many top colleges and universities across the country. In 2005 we did $8 million in sales. In 2020 we expect $40 million in total sales across our enterprise. Yes, this only makes us a medium-size company, but we pack a lot of punch for our size—as our client list suggests.

I can't promise the kind of HR and PR benefits your company will get by making a commitment to sustainability, clean energy, and green buildings, but what I can say is that doing the right thing more often than not allows you to reap these strategic advantages. Don't let the bean counters talk you out of leading.

The best part of our story is that our company is just a microcosm of thousands of businesses, schools, and government entities across the United States and around the world that are also reaping these strategic advantages. And, cumulatively, we represent the dawning of the clean energy revolution. Why is this important for conservatives to acknowledge and understand? Because it is fast becoming the greatest economic, security, health, and environmental opportunity of the twenty-first century.

Oh, and about Mary Frances … her initial hesitation about my investing our relatively meager but hard-earned savings in a vision that promotes a better future for our children and grandchildren has developed into a spiritual belief, and even a conviction. "Steve, God has so blessed us through your work and trust in him," she often says today. Her words inspire me all the more.

PART 2
THE PROBLEM
IS FOSSIL FUELS

CHAPTER 4

We Pay for It with Our Health and Environment

t's probably time for me to take a step back here and address some questions you might have about this story I'm telling. The principle objection I expect, because I've certainly heard it before, from friends and … "not exactly friends," is this:

"Great that you were inspired by the clean energy and green building movement, Steve. Took a chance on bucking the status quo and investing in an LEED-certified building, and it paid off in so many ways, hoped for and unexpected. But my life is unfolding nicely in the fossil fuel economy. I like driving the car I have, I have a good job, my priorities are my own, and the world of solar panels and wind farms and electric cars … that's just not what I'm worried about. If that's the future, great. My kids will drive electric cars and stick solar panels on their rooftops."

And there's another argument to be made, too, one I absolutely sympathize with:

"The energy economy we have has been pretty good to this country, and I don't want to change my way of life or disrupt the livelihoods of hardworking people who make their living in the fossil fuel industry."

Believe me, I get those points and many others made by friends, family, team members, and leaders in our community. We have an energy system that more or less works to power a great economy and happy lives; why mess with that? But here's the problem.

This is not just a neutral choice like "You like tennis; I'm into baseball." There are some serious problems with our fossil fuel economy that we all pay for, regardless of opinion, party, race, religion, or nationality. Many of those costs are hidden to those who aren't focused on them, but they are very real, and they are becoming catastrophically high. "Business as usual" has run out of time as a plausible option. The simple facts are these. The energy that we extract from fossil fuels is a hazard to our health. Maintaining a steady supply of oil, gas, and coal carries a much higher cost in American lives and treasure than most people realize. Our dependence on foreign fuel imports often pushes us to betray the ideals we hold dear. And perhaps the biggest problem with fossil fuels is that the battle against clean energy innovation and a brighter future is not being fought fairly. There are few bigger fans of capitalism than I am, but the continued dominance of Big Oil is not an example of the free market choosing winners. The game is rigged, and all of us, along with our children, are paying for it.

If I were to ask you to name a few negative impacts of fossil fuels, the first things that would come to mind would likely be impacts related to burning fossil fuels—power plant emissions, the pollutants coming out of the tailpipes of our cars, factories, airplanes. But the damage caused by fossil fuels begins when we take them out of the ground. Coal mining is a dangerous job, with a constant risk of instant death by accident or slow death by black lung disease. In the Appalachians, hundreds of mountains have had their forests clear-cut and their tops blasted off.[1] Debris fills the valleys, and mine waste fouls the streams.

Oil extraction also poses dangers. Leaks from underwater wells, barges, tankers, and pipelines can devastate beaches, fish, and wildlife along with the economic well-being of whole towns, cities, and states. One need only remember the Exxon Valdez oil tanker spill in 1989 and the BP Deepwater Horizon oil rig explosion and leak in 2010 to understand the destruction that millions of gallons of oil can have on people, planet, and profits.

Pollutants released in oil drilling can also cause irritation of the eyes and throat, blistering, and nausea—and that's in the short term. With long-term exposure, they can lead to blood disorders, cancer, birth defects, numbness, blurred vision, and muscle weakness.[2]

Fracking has led to the rapid growth of US oil and natural gas production but has also polluted groundwater, choked rural areas with smog, and even caused earthquakes. Oklahoma used to have one or two magnitude 3.0 earthquakes per year. Now it has one or two per day—more than California[3]—and the state says "the majority of recent earthquakes in central and north-central Oklahoma are very likely triggered by the injection of produced water in disposal wells" related to fracking.[4]

Once oil, coal, and gas are removed from the earth, the threats to our health and safety only multiply, simply because these fuels are toxic and unstable by nature. Natural gas, for example, remains explosive even once piped to its final destination and delivered for use in homes, schools, and businesses. And not only is natural gas toxic and explosive, much of it is leaking into our air. A 2018 study for the journal *Science* spearheaded by the University of Colorado found that EPA estimates of a 1.4 percent rate of methane loss from natural gas is actually closer to 2.3 percent, which translates to thirteen million metric tons of methane released into the atmosphere from leaks each year, virtually erasing any benefit from replacing coal-burning plants with natural gas.[5] And that is also thirteen million metric tons of toxic and explosive methane in our homes, schools, and businesses. We are so accustomed to media reports of discrete incidents of gas explosions that we have come to accept that exploding buildings and gas fires and damaged pipes that leak toxic gases into the air are just an inevitable cost of powering our economy and lives.

My oldest daughter, Monica, has a close friend who lived in an apartment in Ohio and noticed a funny odor one day. She was unfamiliar with the smell of natural gas and thought she could counter the odor by lighting an aroma candle. When the gas filling her apartment ignited, she suffered first-, second-, and third-degree burns over much of her body. It's a far-from-uncommon story. Similar natural gas explosions happen more often than we realize—throughout our country and the world.

Coal-fired power plants are also fraught with significant costs and risks to man and nature. Pollutants they emit are linked to asthma attacks and chronic lung disease.[6] There is evidence power plants are also linked with neurological problems such as reduced cognitive performance.[7] In fact, in the Ohio Valley (Ohio, Indiana, Illinois, West Virginia, Pennsylvania, and Kentucky), where coal is the dominant source of electricity, there is a high incidence of asthma and related ailments.

Ohio happens to be my home, and the problem of heightened

respiratory illness is a particularly personal one for our family. I will never forget being awakened by Monica in the middle of the night when she was about five years old. She was gasping for air and said she couldn't breathe. I don't have words to convey the panic and helplessness my wife and I felt. Monica was diagnosed with chronic acute asthma, and that first trip to the emergency room became distressingly common, and breathing treatments, puffers, and daily medications were a way of life. As a young adult, Monica still depends on her medicines to get through the day. My youngest daughter, Laura, suffers from asthma, too, and has to take medicine every day to cope.

Nationally, asthma affects more than seven million children under the age of eighteen. It is the third-leading cause of hospitalization among children under age fifteen and one of the leading causes of school absenteeism.[8] Think of the resulting drag on our society and quality of life. The $50 billion plus in annual healthcare costs. The millions of lost school days. And the $6 billion lost in productivity every year from parents who need to care for their children instead of going to work.[9]

And fossil fuels aren't just adding toxins to the air we breathe but also often pollute our water supplies. Ash, a byproduct of burning coal, contains toxic heavy metals and leaches into groundwater from the unlined pits where it's stored at some power plants. A Duke Energy plant in North Carolina spilled up to thirty-nine thousand tons of the stuff in 2014, fouling the Dan River.[10] This came after a 2008 spill in Tennessee let loose more than one hundred times that amount—1.1 billion gallons.[11]

There are so many toxic side effects that come as the unavoidable price of extracting, processing, delivering, and burning fossil fuels. But of course humans are not the only ones paying the price. Every living thing on the planet, in fact our planet itself, suffers from the toxic effects of our fossil fuel–powered economy. And as bad as all of those effects are, the most dangerous byproduct of fossil fuels might be one we can't see or smell. The very gas we exhale—carbon dioxide—is the environment's public enemy number one. Why? Unlike the local and regional pollution other chemicals can cause, the effects of carbon dioxide are global and last for centuries.

Most of us know about the greenhouse effect, whereby carbon dioxide and other gases trap heat from the sun in Earth's atmosphere.[i] Carbon dioxide occurs naturally, of course, and in fact, in proper balance, it is a necessary component of life on Earth. The problem comes when we disrupt the delicate balance of the atmosphere in which the planet's species—and our civilization—has evolved.

Fossil fuel is a general term for buried or trapped geologic deposits of organic materials, formed from decayed plants and animals that have been converted to crude oil, coal, natural gas, or heavy oils through exposure to heat and pressure over hundreds of millions of years. This organic material is full of trapped energy that can be burned for fuel, just as you might burn a log in your fireplace to produce flame and heat. The difference is that while burning wood releases carbon that had been in the air over the lifetime of the tree, the carbon in fossil fuels has been buried for eons—and we've been setting it free at an alarming rate in the span of a few centuries, since the Industrial Revolution began in the 1700s.

Plant life and soil naturally absorb and store carbon from the atmosphere, but Earth's dwindling forests and farmlands can only absorb a fraction of the carbon that is now being released through human activity and the effects of global warming. Every day, according to environmen-

---

i   Some other gases, such as methane, trap even more heat than carbon dioxide, but carbon is more worrisome because it's far more abundant and stays in the atmosphere much longer.

talist and entrepreneur Paul Hawken, human beings burn "an amount of energy the planet required 10,000 days to create."[12]

That's why the concentration of carbon dioxide in the atmosphere is at its highest level in at least eight hundred thousand years, according to the Intergovernmental Panel on Climate Change (IPCC), a United Nations–organized group that reviews and assesses climate-related science.[13]

Global warming skeptics will point out that the concentration of carbon dioxide in the atmosphere changes naturally over time. But what we're seeing is unprecedented. Here are measurements of atmospheric carbon during the past eight hundred thousand years:

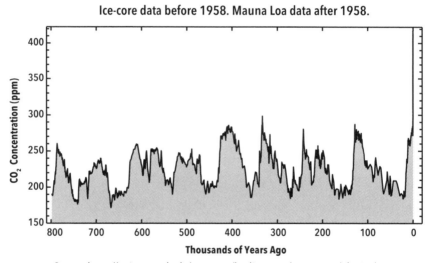

Source: https://scripps.ucsd.edu/programs/keelingcurve/wp-content/plugins/
sio-bluemoon/graphs/co2_800k.pdf

At first glance, you might assume the black line at the right of the chart is the chart's border, since it seems to be a perfectly straight vertical line. Look again. The fact is, that line shows carbon dioxide concentrations in our atmosphere suddenly more than doubling over historic lows and growing to 30 percent higher than any recorded levels over the past eight hundred thousand years. We are running an uncontrolled experiment on our planet, one that can't easily be reversed. The concentration of carbon dioxide in the atmosphere was at less than three hundred parts per million (ppm) at the start of the industrial era. We are now at over four hundred ppm. With the gases we continue to pump into the atmosphere, by most projections, we will top one thousand ppm within a century.[14]

> **We are running an uncontrolled experiment on our planet, one that can't easily be reversed.**

It is no coincidence the past thirty years have been the warmest period in at least eight hundred years.[15] The ten hottest years on record have come since 2005, and the five hottest years on record have been the five consecutive years prior to the writing of this book: 2015, 2016, 2017, 2018 and 2019.[16]

It takes an enormous amount of heat energy to cause any rise in Earth's average yearly temperature, but that figure has risen by a full two degrees since the preindustrial era. That might seem like a negligible change, but it means there has been a massive increase in accumulated heat in our atmosphere. That extra heat is driving regional and seasonal temperature extremes, reducing snow cover and sea ice, intensifying extreme weather, and changing habitat ranges for plants and animals—expanding some and shrinking others.[17]

Let me come back to the point I made earlier: this is not a neutral choice. "I'm into renewables; you like fossil fuels … "traditional" energy … the way it's always been." As it turns out, the way it's always been is not at all how it's always been. Moreover, it is not our only option, nor can it be our way forward.

The cost to our health and our environment is one we can simply no longer afford to pay.

CHAPTER 5

We Pay for It with Hidden Taxes

ere's a simple question that many of us have never really asked: When did we begin to use fossil fuels as a source of energy?

Well, to be accurate, the answer is "for most of human history." Ancient civilizations used the petroleum that they found seeping onto Earth's surface for building and waterproofing. The Chinese found oil reservoirs in salt wells and drilled wells as deep as one hundred feet to reach natural gas and oil around 500 BC. There is evidence that a few hundred years later, the Chinese had developed to the point that they were using bamboo pipes to carry natural gas to homes for heating and lighting.

But it wasn't until the Industrial Revolution in Britain, which began in the middle of the eighteenth century, that coal and later oil and natural gas became key energy sources for industry and house-holds in Britain and eventually other industrialized countries. With James Watt's invention of the steam engine, powered by coal, the era of large-scale consumption of fossil fuels began. In the years that followed, coal, oil, and later natural gas became the principal source of power for machines, manufacturing production, and factories that built the modern era.

At the end of the nineteenth century, a new invention would accelerate the demand for petroleum products around the world forever—the first automobile with an internal combustion engine, invented by German engineer Carl Benz in 1885. The automobile was first mass-produced in Michigan by Henry Ford, and by 1909, oil production in the United States more than equaled that of the rest of the world combined. It's no wonder that the world's first billionaire, John D. Rockefeller, was an American oilman.

Rockefeller once said that "the best business in the world is a well-run oil company. The second-best business in the world is a badly run oil company." Rockefeller understood that the postindustrial world would rely on energy that was cheap, mobile, and plentiful and that whoever supplied that energy would hold an unassailable economic advantage. And that reality essentially powered the twentieth century and the rise of the United States as the preeminent economic, military, and political power of the era.[18]

Fossil fuel was undeniably one of the core pillars of the ascendency of the United States in the twentieth century. And yet, two decades into the twenty-first century, the energy that is produced by oil and other fossil fuels has long since ceased to be a source of strength for our nation. It's true that in 2018 the United States became the world's largest crude oil producer, ahead of Russia and Saudi Arabia.[19] And you've probably heard a lot about the boom in shale oil production in certain parts of the country, particularly North Dakota. But the United States is also, by far, the world's largest oil consumer.[20] Even with the new production we've seen in the past few years, our nation is still the world's largest net importer of petroleum.[21] Our net imports amount to more than a quarter of the oil we use.[22]

Granted, that's down from a high of 60 percent in 2005. But even with American production at an all-time high, we still consume far more oil than we drill. In 2018 the United States imported about 9.9 million barrels of petroleum per day—using about 19.0 million barrels a day while producing about 11.6 million.[23] Even with record increases in production from shale and other sources, it still isn't enough to balance the ledger. Realistically, we cannot drill our way out of our oil deficit.

As long as we rely on imported oil, every American will pay a hidden "oil tax" that sends twenty-five cents overseas for every dollar we spend on oil. That's money not recirculating in our economy and creating jobs at home. As economists Alan S. Blinder and Jeremy B. Rudd put it, "If imported energy, which mainly means imported oil, becomes more expensive, the real incomes of Americans decline, just as if they were being taxed by a foreign entity. The [oil] 'tax' hits harder the less elastic is the demand for energy, and we know that the short-run price elasticity is low."[24]

In other words, as long as we need to import oil, we are at the mercy of oil-exporting countries—some of which are among the most unstable and dangerous in the world. Every time there is terrorism in Nigeria, civil unrest in Venezuela, or conflict in the Middle East, Americans' wallets suffer collateral damage. Sometimes, this transfer of wealth can be a minor drag on the American economy. At other times, it is a major threat that can plunge the country into a recession.

Consider what happened in the 1970s.

After the 1973 Arab-Israeli war, Arab nations placed an oil embargo on countries supporting Israel. The price of oil doubled in six months, sparking inflation across the US economy. Blinder estimates that extra cost forced Americans to send 1.5 percent of their annual gross domestic product overseas. Not having that wealth circulating in the economy lowered economic production here by about 3 percent—enough to make the difference between healthy growth and stagnation.[25]

Of course, many oil-producing countries are too dependent on exports to keep their oil off the market forever, so these "oil shocks" eventually subside. But they always seem to come back. Consider this chart showing the price of a barrel of crude oil in real dollars since 1947:

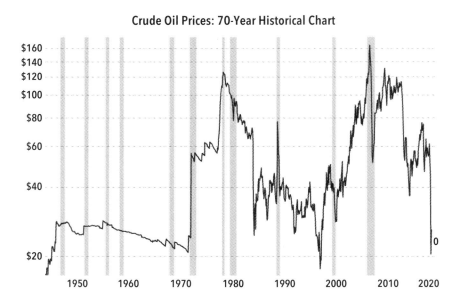

Crude Oil Prices: 70-Year Historical Chart

Source: Charles Schwab Macrotrends,
https://www.macrotrends.net/1369/crude-oil-price-history-chart

If the cost of labor were that volatile, would you hire as many people? If property taxes were that unpredictable, would you expand your offices?

The Arab embargo is only the first of the spikes. There was another after the Iranian revolution in 1979, then another around the 1990–1991 Gulf War. But the first two decades of this century have seen another destabilizing factor caused by the disruption of production. The steady increase in oil prices that began around the year 2000 was not the result of war, embargo, or political upheaval. This steady price rise came from a global increase in demand, especially in rapidly developing countries such as China and India—at a time when the world's ability to produce more oil is getting close to maxing out. Yes, oil prices have plummeted, even dropping below zero for the first time in history, as the sudden global shock of the COVID-19 pandemic caused a collapse in demand and consumption, but once industrial economies go back online, so will their insatiable appetite for oil.

But clearly global economic calamities and new production that requires high prices to be economical are not sustainable ways to bring energy prices under control. Unless we change the way we power our economy and day-to-day lives, higher oil prices will be the long-term trend. America's CEOs should take note: betting the future of their companies on continued cheap oil is a foolish bet.

Consider this chart of US oil use:

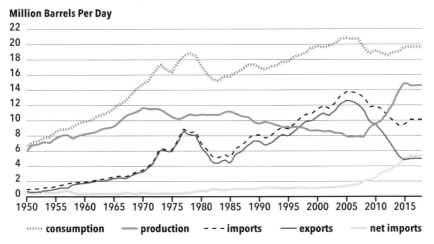

US Petrolium Consumption, Production, Imports, Exports, and Net Imports, 1950–2018

Source: US Energy Information Administration, *Monthly Energy Review*, Table 3.1, April 2019

As you can see, imports are down, but that's mostly because production is higher, not because demand is lower. Now think about where the recent boost in production is coming from. We're drilling in places that were never economical before—shale and tar sands. A shale well's production will drop by about 75 percent in a year, so exploring and drilling is a constant process.[26]

For their next big finds, the oil companies are eager to open up environmentally sensitive areas of the Arctic—both on land and at sea. Why are they going to all that expense? Because the easy oil is running out. Let's not forget: we are talking about a limited natural resource. Even drilling in far-flung places like the Arctic—which is only feasible when prices are high—is a temporary fix at best. What comes after that?

To some extent, we did learn our lesson in the 1970s. Cars became more fuel efficient, and electric utilities have shifted away from totally oil-based power generation. We have proven we can grow the economy without increasing our use of fossil fuels. But we still are a long way from freeing ourselves of the oil tax.

The problem will get worse before it gets better because the upward pressure on oil prices is not going away. The Chinese and Indian economies are projected to keep growing at rates of more than 5 percent per year for years to come. That's more than 2.5 billion people moving ever closer to Western standards of living, with the energy use that requires. Liquid fuel consumption is expected to nearly double in both countries by 2040.[27]

As economist Jordan Hamilton writes, "Already China is the world's biggest market for buying new cars. Even so, China has only one passenger vehicle per thirty residents, compared with one vehicle per 1.3 residents in the United States."[28] China may have a lot of catching up to do, but it's doing it fast. Since oil supplies cannot increase indefinitely, global use will have to be rationed, and the way the market does this is by raising prices.

And these "hidden taxes" on foreign oil are not the only indirect costs you pay. Before you ever pull up to the pump at the local gas station, you've already paid a sizable amount for each gallon of gas in domestic hidden taxes. A recent report by the International Monetary Fund (IMF) found that US spending on direct and indirect subsidies for coal, oil, and gas reached $649 billion in 2015. Total Pentagon military spending for that same year totaled $599 billion. It's a staggering figure that places the United States behind only China, the largest subsidizer of fossil fuels by far, at $1.4 trillion in 2015.[29]

It's true that the IMF study defines the term *subsidy* very broadly to include more than direct payments or tax relief. It accounts for the "differences between actual consumer fuel prices and how much consumers would pay if prices fully reflected supply costs plus the taxes needed to reflect environmental costs," along with healthcare

> **Every step we take away from fossil fuels and toward energy efficiency and clean, free, renewable energy reduces the high price we pay for our addiction and makes our economy more stable.**

costs associated with air pollution, including premature deaths.[30]

Every step we take away from fossil fuels and toward energy efficiency and clean, free, renewable energy reduces the high price we pay for our addiction and makes our economy more stable. Most of these costs are invisible and don't appear as energy subsidy line items in our national budget or at the pump.

But make no mistake, these costs are hidden in the taxes we pay and impart an insidious drag on our nation's prosperity and future economic potential.

CHAPTER 6

We Pay for It with Dictatorships and Terrorism

O n the morning of May 9, 2019, the massive nuclear-powered aircraft carrier *Abraham Lincoln* and its entire strike group passed along the Suez Canal through the heart of the Sinai Desert, rising out of the sand like a surreal mirage. This deployment of daunting US firepower was making its way from the Mediterranean to the Persian Gulf in response to Iranian attacks on commercial oil tankers two weeks earlier. Despite the seeming disparity in force between the overwhelming firepower of the US force and Iran's naval capabilities, it was a dangerous scenario. Every crew member on those ships knew that the Iranians had developed highly capable asymmetric warfare capability, focusing on small, fast, and less expensive attacks like mines placed by special forces, "swarm attacks" by high-speed gunboats, torpedo attacks from small, stealthy submarines, and cruise missile barrage attacks targeting a single ship.[31]

It is dangerous, stressful, and exhausting duty for our sailors and comes at a massive cost in military spending, potential for rapid escalation of conflict, and the untold costs to military men and their families. This particular deployment by the *Abraham Lincoln* strike group broke all records, keeping the crew from their home port in San Diego for ten months, a post–Cold War record. And our sailors were there, in harm's way, in hostile territory on the other side of the world, principally to protect the world's oil supply.

It would be bad enough if the worst damage caused by our dependence on fossil fuels was to our pocketbooks. But the constant string of deployments by US forces around the world to secure and protect oil supplies is evidence of an even greater cost than the instability of our financial markets, hidden corporate subsidies, and higher prices to heat our homes and fuel our vehicles. That cost shows up in the US defense budget and in the impact our fossil fuel dependency has on freedom and human advancement around the world. It's not just our economic vitality that is threatened; it is our safety, our moral leadership, and the bedrock principles upon which the United States was founded.

Our dependence on oil, and the remote and often turbulent locations where much of it is produced, means the US military has effectively become the oil companies' global security force, and we are paying for it in dollars and lives. The cost of fossil fuel energy to our national security will only get worse if the world maintains its current unsustainable dependence on fossil fuels to power our future. Even if the United States could produce all the oil it uses, our navy would still be required to safeguard the movement of oil for the rest of the world. Major energy producers—many of them autocratic states such as Russia and Saudi Arabia—will continue to hold immense diplomatic leverage, shielding them from pressure to make democratic reforms.

WE CAN'T QUIT THE MIDDLE EAST

You don't have to be a pacifist to note the immense toll military entanglements in the Middle East have taken on our country. More than 4,400 Americans died during the Iraq war. Estimates of the cost of that war vary widely, but it surely will top $1 *trillion*—all of it added to the national debt, much of it borrowed from China.

In 2000 the attack on the USS *Cole* in Yemen killed seventeen sailors, and our retaliatory drone attacks in that country killed dozens of civilians along with their terrorist targets. Twice in the past thirty years, US planes have bombed Libya. And more recently, while the *Lincoln* strike group faced off with the Iranians in the Strait of Hormuz, our air and land troops were back in Iraq fighting the terrifying militants of ISIS.

I am not arguing that oil is the only reason for our military involvement in the Middle East, but no serious argument can be made that maintaining the supplies of fossil fuel is not chief among them. Oil prices are set in the global market, and a disruption in oil supply anywhere in the world means higher prices here at home. Essentially, the health of our economy is tied directly to the health of some of the most unstable regimes in the world. As long as we remain dependent on oil, we remain tied to those volatile regions.

It doesn't matter whether the gasoline in our cars comes from Russian or Canadian oil. When we buy oil, we create demand. And that drives up prices, which helps all oil producers. Oil is fungible. If we buy from Norway or Mexico, it doesn't mean the Saudis lose. It means they sell to China or Europe instead. We can't do business with our friends without also funding our foes. Countries benefiting from the world's fossil fuel addiction are very often the ones contributing most to the extremist ideologies that lead to terrorism.

In a memo aired by WikiLeaks, then secretary of state Hillary Rodham Clinton complained that Saudi Arabia was reluctant to stop the flow of money from its citizens to terrorist groups, including al-Qaeda and the Taliban. Other key sources of funds, the 2009 memo stated, were Qatar, Kuwait, and the United Arab Emirates.[32] Those countries are four of the eighteen largest oil producers in the world—and nominal US allies. Oil wealth flows through these key financiers of terrorism to the very attacks that the United States and government security forces around the world spend billions to thwart. A bombing in a metro station in Paris, a roadside bomb detonating in Fallujah, the drone rocket attacks that sent over seven thousand US sailors into harm's way as part of the *Lincoln* strike group—this never-ending litany of bloodshed and conflict is funded, and necessitated, for the most part by oil.

Playing ball with authoritarian governments ensures the flow of oil, but at the cost of alienating the oppressed people of those countries. From the perspective of many in the Middle East, US involvement there makes Americans look "imperialist, power-hungry, oil-seeking, and crusading."[33] That perception only plays into the hands of al-Qaeda and other dangerous terrorist groups. It is further fueled by Wahhabism, the extreme religious ideology Saudi oil money has promoted around the Muslim world.

Another fossil fuel producer whose bad behavior has been tolerated for too long is Russia. Much of Europe is dependent on Russian natural gas for heat in the winter, and President Vladimir

Putin has not been afraid to use the leverage this offers him.

Because about half of the Russian natural gas going to Europe passes through Ukraine, this adds the Russia-Ukraine feud to the list of conflicts that have ripple effects across world energy markets. When Russia cut off supplies to Ukraine, either for political reasons or in a dispute over prices, Ukraine siphoned off gas from the pipeline for its own use starting in 2005. In 2009 Russia responded by cutting off supplies to the European Union for two weeks. There is a limit to how long Russia can maintain an oil and gas export shutdown because it is dependent on oil and gas revenues, but it has already benefited from the high-stakes game of chicken it is willing to play.

The 2009 shutoff spurred Russia to construct a new pipeline that bypasses Ukraine, weakening that country's leverage in price negotiations.[34] It also demonstrated Europe's energy dependence, which makes the European Union reluctant to impose tougher sanctions on Russia for its war in Ukraine. It's a form of economic and security extortion that forces democratic nations to weigh principle against the realities of their energy economies. Even if European countries have sometimes been guilty of bending their principles to such pressure, on this subject the United States is in no position to throw stones.

From Iran to Venezuela, oil money has allowed autocratic regimes to stay in power—often for decades—while refusing to make reforms that would empower their people and unleash their economies but also jeopardize their control, and this isn't a random pattern. Economists have noted what appears to be a "resource curse," a pattern where countries rich in oil, minerals, or other natural resources are often unable to provide political stability, economic growth, and a rising standard of living to their people. As hard as it is to imagine a country being too dependent on free, sustainable energy, economies wholly dependent on fossil fuel production have generally been associated with oppression, damage to the environment and public health, and extreme resource inequity.

There is debate about the reasons for this, but there is no shortage of candidates. Oil production is capital intensive, so it produces relatively few jobs and concentrates wealth in a small number of hands. A political class that controls resource wealth may have less incentive to invest in education and industrialization. Because oil states are funded by their natural resources rather than by taxing their citizens, they have less need to seek legitimacy through elections. Countries that are dependent on the unpredictable swings of commodity prices can also be subject to economic instability and accompanying political turmoil.

Whatever causes the problems of oil-dependent nations, they can't be confined within those countries' borders. Rivalries for control of these resources can lead to armed conflict, which prevents economic growth and can cause refugee crises and fighting that spreads across regions. And it's clear that the more the developing world depends on oil, the less stable and prosperous it will be.

For an admittedly extreme example, consider Japan in the 1930s—when its need for resources led to imperialism and ultimately World War II in the Pacific. But that ugly history doesn't seem so remote when one considers the current diplomatic and military tussles between China and its neighbors over various uninhabited islands in the South China Sea. Why would major powers engage in saber-rattling over empty rock piles? Again, it's a familiar song—for the fossil fuels that potentially could be extracted offshore. If the world doesn't reduce its thirst for oil, emerging nations and multinational oil companies will have to go ever farther to find it. And the potential for conflict becomes ever greater.

A TARGET AT HOME AND ABROAD

More than forty people were killed by al-Qaeda–linked terrorists at a natural gas facility in Algeria in 2013. That year there were nearly six hundred attacks on oil and gas infrastructure and personnel—an all-time high, and nearly a quarter of all the terrorist attacks worldwide.[35] The total for 2014 rose even higher thanks to ISIS, which, at the peak of its ascendance, was making up to $50 million a month from the oil fields under its control.[36]

Meanwhile, if you think even under the American military umbrella the threat to energy infrastructure is contained to distant parts of the world, consider the fact that US intelligence agencies have revealed that countries like Russia, China, and Iran have already either attacked foreign grids or conducted reconnaissance on the US grid,[37] and a recent report from the Government Accountability Office said cyber threats to US critical infrastructure like the energy sector are increasing.[38]

"When you talk about cyberattacks against the energy infrastructure, primarily you are looking at nation states like Russia, China, Iran," said Caitlin Durkovich, who served as DHS assistant secretary for infrastructure protection during the Obama administration. "They may come at it with different motives, but certainly they have built up a significant capability."[39]

Foreign actors and state security services have also been implicated in cyberattacks on US oil and gas pipelines. A recent ransomware attack caused a US natural gas compressor facility to shut for two days, the latest in a string of attacks targeting American energy infrastructure over the past few years. According to the Department of Homeland Security, the attackers used an email phishing attack to gain control of the facility's information technology system then identified critical assets and disabled security processes within the system before making a ransom demand.[40]

According to a National Academy of Sciences report, a well-planned attack "could deny large regions of the country access to bulk system power for weeks or even months."[41] The potential effects of that are almost too terrible to contemplate: civil unrest, economic production crippled, and possibly the deaths of hundreds or even thousands.

But what's the connection, you might be asking yourself, between attacks against national energy grids and a fossil fuel energy economy? Why would a clean energy grid be any less vulnerable to terrorism or cyberattack than a power grid fueled by fossil fuels? The answer is decentralization, or microgrids. Our electric grid will remain vulnerable as long as it depends on centralized power plants distributing power to millions of homes and businesses. Only a system with tens of thousands of small, interconnected solar, wind, and geothermal generation sites eliminates the possibility of an attack that would knock out power in large sections of the country. When our vehicles and buildings produce the energy they consume, the threat of wide-scale interruption is disrupted.

No one knows the vulnerability of our energy system better than the United States military, which has learned through hard experience in Iraq and Afghanistan. Running vehicles, computer equipment, air conditioners, and everything else a modern fighting force needs in rugged, isolated terrain requires enormous amounts of fuel, and transporting it to the front lines is costly and dangerous. That's why the cost of our military deployment in Afghanistan since 2001 is approaching $1 trillion two decades later,[42] and fueling those operations costs taxpayers an average of *$400 per gallon.*[43]

It is not clear how many lives have been lost not only protecting international oil supplies but simply serving the military's energy needs, but most casualties in Iraq were caused by roadside bombs, many targeting fuel supplies—sadly, inviting targets for the enemy.

Eventually, the military began replacing some of its fuel shipments with solar panels and energy-saving lights—for its own use and to help local communities develop. Their reasons had nothing to do with concerns about the environment. It was about saving lives and dollars.

The military also knows that a healthy environment is crucial to our national security. That's why the Department of Defense has released a Climate Change Adaptation Roadmap, which examines ways to deal with the harm we have already caused. According to the plan, "rising global temperatures, changing precipitation patterns, climbing sea levels, and more extreme weather events will intensify the challenges of global instability, hunger, poverty, and conflict. They will likely lead to food and water shortages, pandemic disease, disputes over refugees and resources, and destruction by natural disasters in regions across the globe."[44]

We'll examine climate change and just how unpredictable its effects will be in the following chapters. And that's exactly the problem for national security planners. We know sea levels are likely to rise by one to three feet over the course of this century.[45] Meanwhile, in Bangladesh, 140 million people live in a floodplain.[46] The potential exists for a refugee crisis that would strain any country, and this is in a developing region where neighbors have a history of conflict—and nuclear weapons. Even if we set aside the humanitarian concerns, it would be impossible for the United States to isolate itself from turmoil on a potentially massive scale.

The problem of climate change is not one for the distant future. It is already here. In fact, we might already have seen our first climate-related war in Syria. Starting in 2007 a severe drought forced farmers off their land into the cities, which were already flooded with refugees from the 2003 Iraq war. The Syrian government was unable or unwilling to respond to the humanitarian crisis. Discontent led to violent protests, followed by a brutal crackdown and eventually civil war. Out of that war came the terrorist group ISIS, which emerged as a bigger threat to the United States and world security than even al-Qaeda.

Of course, drought was not the only cause of the war. And severe weather can happen with or without a changing climate. But researchers say man-made climate change made the unprecedented drought in Syria two to three times more likely than it otherwise would have been.[47] When climate change destroys people's livelihoods, it is not surprising when they respond with violence. Not all threats to national security are from foreign militaries. The global environmental problems caused by the United States and the rest of the industrialized world inevitably come back to haunt us at home.

> The ability to remove energy from the equation in our diplomacy would allow us to focus on other priorities—nuclear nonproliferation, promoting human rights, and addressing current and future disease pandemics.

That's why a panel of eleven three- and four-star generals and admirals has called for the United States to reduce its oil consumption by 30 percent.[48] A reduction of that size, they argue, would insulate the United States from the effects of a supply disruption—meaning we would no longer be at the mercy of oil-rich enemies.

The ability to remove energy from the equation in our diplomacy would allow us to focus on other priorities—nuclear nonproliferation, promoting human rights, and addressing current and future disease pandemics. Moreover, the technologies that would replace oil would make our military more nimble, mobile, and better able to keep us safe at a reasonable cost.

We are fortunate to have military leaders who know that being a greener country will make us a stronger country. And being stronger means being safer from threats of all kinds—the caustic influence that autocratic regimes can have on our national values, those people who would intentionally harm us, as well as potentially deadlier dangers that would come from drastically changing our environment.

CHAPTER 7

We Pay for It
Because of Greed
and Corruption

N ot long after the grand opening of our LEED Gold–certified headquarters in Milford, I met my friend Mark for breakfast at Coffee Please, a local favorite. I was happily telling him all about our new building, the excitement and engagement of our employees, and the growing interest from our customers because of all the good PR we were getting.

"That's phenomenal, Steve; it really is," Mark said. "But here's what I don't understand." He leaned back and clasped his hands behind his head. "The building more than pays for itself, your employees are happier and more productive than ever, and business is booming because your clients approve ... so why isn't every business doing this? For that matter, why isn't every home built this way?"

Your mouth to God's ears, Mark, I thought. *Why indeed?*

Mark was asking the obvious question, but it's not necessarily as simple a question as it seems. And the answer is certainly not simple.

Americans, particularly those of us on the conservative end of the ideological spectrum, have historically put great faith in the "invisible hand" of the free market first articulated by the Scottish economist Adam Smith. Smith promoted a laissez-faire economic model based on the belief that a constant interplay between market supply and demand causes the natural movement of prices and the direction of trade. Put another way, the "invisible" hand of supply and demand will naturally push markets to equilibrium without government interventions choosing winners or losers. In essence, Mark's simple question exposed him as a disciple of Adam Smith. In Smithian terms, he might have asked, "If renewable energy sources make economic sense, fill a need, and provide energy with a superior cost-benefit ratio, why aren't they being widely adopted as part of markets moving naturally toward equilibrium?"

I'll acknowledge there are plenty of impediments to revolutionizing the energy economy of a nation, let alone the world. These challenges include the lack of a coherent national strategy; a tangle of local, state, and federal regulations; regional differences in solar, wind, and geothermal resources; and the obvious time and investment needed to accelerate the growth of an electricity generation and distribution sector while powering down another. Workers must be trained to build and operate new equipment, battery and solar cell research must be ramped up, and America's clean energy future must be built out, house by house and business by business.

But what is slowing the adoption of clean energy in America is not simply the invisible hand holding us back because the free market still favors fossil fuels. The fact is that Smith's invisible hand is being tied behind our back by very wealthy, very powerful people who don't want change. These people spend billions of dollars currying favor with politicians at the federal, state, and local levels in order to win their loyalty and influence how they legislate to protect, subsidize, and sustain the energy status quo, even if it puts them out of step with science and the public interest and squanders their own opportunity to apply their expertise and resources to leading the next energy revolution.

The people I'm talking about have built their fortunes in the fossil fuel industry—broadly defined as coal, oil, gas, and utility companies. All these businesses have been deeply intertwined with government since their inception. These folks are not shy about using their influence to sway policy makers, and they are expert at it. We can't talk about the problems with our energy economy without talking about the companies that benefit from the status quo and that are doing their best to perpetuate it.

A common defense of lobbying is a First Amendment defense. Let's remind ourselves: What are our First Amendment protections, exactly? "Congress shall make no law respecting an establishment of religion, or prohibiting the free exercise thereof; or abridging the freedom of speech, or of the press; or the right of the people peaceably to assemble, and to petition the Government for a redress of grievances."[49]

So what we're talking about here, in regard to lobbying, is the right "to petition the Government for a redress of grievances." In short, industry lobbyists have the right to speak to their elected representatives and make their case, and others who disagree have the same right. There's a simple logic and sense of fairness to the argument, but again, like Smith's laissez-faire theory of markets and equilibrium, it assumes that those arguments are operating on a level playing field, that lobbyists are simply arguing for their point of view, and that their representatives are under no undue influence to accept or reject that point of view beyond their best analysis of the equity of the case at hand.

A level playing field—the constitutional right to simply be heard and state your case to your representatives. Nothing could be simpler, more essential to democracy, more American. But the fact is, the fossil fuel lobby has essentially acknowledged, through their actions if not their words, that they can't win this argument in Adam Smith's neutral arena. And though they lean on the First Amendment to protect their right to lobby, they are certainly not interested in a level playing field.

Get ready for some big numbers—staggering, really.

The OpenSecrets database is a project sponsored by the Center for Responsive Politics whose stated mission is to provide "access to clear and unbiased information about money's role in politics and policy and to use that knowledge to strengthen our democracy."[50] According to publicly available information compiled by their researchers, in the 2017–2018 midterm election cycle, corporations, individuals, and trade groups in the fossil fuel industry spent $265,773,915 on lobbying and made $93,392,002 in contributions to national-level candidates, parties, and outside groups. That means that, over a two-year period, the industry spent more than $359 million, or nearly $500,000 per day, on lobbying.

How did the other side of this First Amendment "conversation," the renewable energy lobby, compare? During the same period, renewable energy companies—wind, solar, and hydroelectric interests—spent $26 million, or just over $28,000 per day. That's roughly a fourteen-to one disparity, which is the dictionary definition of an "unlevel playing field."

And these figures cited by the OpenSecrets research reflect only "persuasion money" spent by fossil fuel interests at the federal level, but oil, gas, and coal interests are just as active at the state level. For example, on recent statewide ballot initiatives, the industry spent $31 million in Arizona and $30 million in Washington to defeat clean energy measures.[51] These are impressive numbers, but they almost certainly underrepresent the real money being spent to prop up fossil fuel energy. More and more of the money going to climate change denial, since the Citizens United decision, has been untraceable—coming through nonprofits that don't have to disclose their donors.[52] So when Exxon or any company says it is no longer funding this or that dubious political cause, the claim is impossible to verify.

But the most telling part of this story isn't just the vastly disproportionate influence spending; it's the fact that the majority of fossil fuel spending is not allocated toward advocating for the gas, coal, and petroleum industries at all. Instead, rather than making the case for continued support of fossil fuels, the industry has shifted to lobbying against clean energy. They have seen the future, and they are working hard and spending freely to delay its arrival. They are no longer trying to win this argument; they have pivoted to distraction, pumping money toward decision-makers to stall action, and actively promoting climate change denial in the public through the airwaves, mailings, social media campaigns, and astroturf organizations.

We've all become familiar with disinformation campaigns on social media or oil company commercials touting their investments in "clean energy." But you may not be familiar with that last one, astroturf organizations, and that's no accident. They maintain the lowest of profiles by design, and they're worth special mention because they illustrate just how opaque, indirect, and downright dishonest fossil fuel's influence campaign against clean, plentiful, and renewable energy has become.

Astroturf organizations are contrived consumer groups, financed by fossil fuel interests, that pose as grassroots public interest groups concerned with a popular cause, with the actual purpose of undermining attempts by government or industry to address the effects of climate change and promote clean energy technologies. And there's the joke: AstroTurf—fake grass, no roots.

The top lobbyist for the fossil fuel industry in the western United States, the Western States Petroleum Association (WSPA), has a member list that includes major fossil fuel players including BP, Shell, ExxonMobil, Chevron, and Occidental. The WSPA's strategy in the western states, confirmed by a leaked 2014 presentation by WSPA president Catherine Reheis-Boydle, was to "use these fabricated organizations to falsely represent grassroots opposition to forward-looking policy on climate change and clean technologies."[53]

The tactic is as dishonest and absurd as it sounds, and yet the disinformation campaigns organized and executed by WSPA astroturf fronts like Washington Consumers for Sound Fuel Policy, Fed Up at the Pump, and the California Drivers Alliance have been remarkably successful. In California, for example, Fed Up at the Pump and the California Drivers Alliance were instrumental in delaying a proposal to place transportation fuels under the state's carbon cap. The following year, the California Drivers Alliance ran an aggressive and ultimately successful campaign to kill a provision in a bill that would have required a 50 percent cut in oil use in California by 2030.[54]

So if it's puzzled you how so many people in the United States are agnostic about whether humans really are changing the climate while anywhere from 97 to 100 percent of publishing scientists concur that they are, it's not a mystery. It's not because there is any meaningful scientific debate about whether or not Earth is warming, nor about the central role played by human activities; it's because fossil fuel interests have spent a lot of money to sow doubt about the science in every possible way. That concerted effort has led the news media to cover the issue as though it were a scientific controversy long after that ceased to be the case. Climate change denial is practically a cottage industry, and a very successful one at that. By a count taken in April of 2017, 232 out of 435 representatives in the House and 53 out of 100 senators deny the existence of human-made climate change.[55]

I saw the influence of the fossil fuel interests firsthand—and how politics can get in the way of the public good—during a frustrating fight here in Ohio. In 2008 the state legislature passed a renewable portfolio standard (RPS) for utilities, requiring them to gradually increase their use of renewable energy to 12.5 percent of their total generation by 2025. At the time, it passed nearly unanimously—and it should be mentioned that more than thirty states have adopted such forward-thinking legislation in recent years.

But after the 2010 Ohio election, things changed. With strong allies now in power, the utilities and their front groups pushed for what eventually became Senate Bill 310, which would make Ohio the first state to move the clock backward and put its renewable energy requirements on hold.

In 2014 I was invited to the state capitol in Columbus to testify before the Senate Public Utilities Committee on the bill. There were hundreds of people in the hearing room, but most of them were utilities executives, lawyers, and lobbyists—all there in a show of strength. And, in my opinion, a show of expectation to those on the committee that their campaign contributions would not be for naught.

SB310 passed, to Ohio's shame. What had changed in only six years? Certainly not the science of climate change, which had grown more convincing. Not the cost of renewable power, which had gone down. One answer is fracking. Drilling companies wanted to produce natural gas in the state, and the governor wanted to promote the practice to help the state's budget. But the other answer is a pay-for-play culture that came back into vogue with the last election cycle. In other words, lobbying gone mad and politicians gone corrupt.

NOT JUST THE PRODUCERS WHO ARE RESISTING, ALSO THE DISTRIBUTORS

Clearly, the massive fossil fuel industry has a lot to lose in the transition to clean and renewable energy sources. But they aren't alone. Energy distribution systems around the world have been built around fossil fuels. Energy utilities also feel threatened, as though their business models are under attack. And in my experience, utility companies have been as aggressive and ruthless in their resistance to clean energy as the fossil fuel industry. A few CEOs have talked a good game about changing in response to sustainability concerns. But in most cases, the rhetoric hasn't been followed by concrete action.

> Not since Thomas Edison has the electric power industry had a visionary leader who wanted to beat the competition rather than prevent competition from taking place.

In my business I've had occasion to meet quite a few high-ranking executives from utility companies. They always seem to be lawyers. Nobody I've met came up through the engineering or marketing ranks, or any other aspect of the business. I believe this partly explains why they are so afraid of competing and slow to innovate. They literally don't know how to do it. Not since Thomas Edison has the electric power industry had a visionary leader who wanted to beat the competition rather than prevent competition from taking place.

What's remarkable is that the utilities know they have a problem, yet until recently most have refused to pivot toward the clean energy future that is obviously coming. For the first time in their history, the utilities are being challenged by a cost-effective competitor—distributed renewable energy. The proportion of American electricity being generated by solar and wind is still relatively small, but it is growing so rapidly and becoming so attractive on price that it seems inevitable that utilities will hemorrhage customers in coming years if they don't pivot from resisting change to leading the transition to clean energy.

At the same time, environmental concerns and rising costs are going to drive continued investment in energy efficiency—further cutting into demand for kilowatt-hours. If these trends take place at a time of low natural gas prices, what happens when fuel costs inevitably rise? Distributed power will look like an even better investment.

Add up all the factors in play, and you have a classic death spiral. The utilities are left trying to recoup their fixed costs from a smaller customer base. That means prices rise and more people becoming their own power generators, which leads to higher prices for those remaining. And the cycle continues, with utility profits falling all the while.

REALLY ONLY TWO PATHS FORWARD

While the transition from fossil fuels to clean, renewable energy may be one of the most fundamental economic and sociological upheavals since the Industrial Revolution, it is certainly not unique by nature. Thousands of industries like retail, music distribution, and advertising have been forced to reinvent themselves in the age of silicon chips and the internet. In fact, it would be far more difficult to list industries that haven't been forced to reinvent the way they do business over the past thirty years.

There are essentially two directions an established industry can go when faced with an upheaval that changes some of the principal underpinnings of the way they operate and generate profits. One is to stick with what has worked before. Often, this also involves actively attempting to undermine whatever new competition is threatening their old model. The other direction is to analyze the new playing field, leverage their established position to change course and win in the new environment, and jettison outdated products, services, and strategies that are no longer valid. One can only imagine that our friend Adam Smith would argue for the latter reaction—to realize that the "invisible hand" is moving their market in a new direction and that they need to reposition themselves toward that new equilibrium.

Examples of both strategies abound. During the twentieth century, the Eastman Kodak Company built a dominant business around photographic film. With the advent of digital photography, Kodak was reticent to transition to a technology that, if successful, would make film, the foundation of its entire business, unnecessary. By the time it became obvious that digital photography would make the majority of film cameras for personal use obsolete, it was too late for Kodak, and it was forced to turn to patent litigation to try to generate revenue. In January 2012 Kodak filed for bankruptcy.

American Telephone and Telegraph is an American company that traces its roots back to 1877, when Alexander Graham Bell invented the telephone and created the Bell Telephone Company. For most of the twentieth century, AT&T dominated long-distance telephone service in the United States and, through its twenty-two regional Bell companies, also controlled most of the country's local telephone service.

But in the 1980s, after AT&T's monopoly was broken up by Congress, new competitors began to offer phone equipment and service, and AT&T's leadership saw the age of personal computers, cellular phones, and the internet arriving on the horizon. Their reaction was quite different from Kodak's. After a failed attempt to enter the computer market, AT&T began to purchase media assets, invested heavily in building a cable network alongside phone lines, and became the largest cable TV provider in the world. At the same time, they built one of the nation's largest wireless networks for cellular phones.

The once-mighty telephone company, in a matter of twenty years, reinvented itself as a media and wireless communication giant. To do it, they had to walk away from a business model that had been very good to them for a century, but they had the foresight to use their assets, brand, and market knowledge to lead the future rather than become a casualty of a new era in communications.

When it comes to our energy economy, the lesson is clear. Utilities need to recognize that the clean energy age is here and embrace distributed, renewable energy rather than cling to their traditional but unsustainable business model. That conversation is definitely happening in utility company boardrooms, and almost every power company is making attempts to say the right things, but in most cases the nation's largest utilities seem to be diverting attention with green talking points and ad campaigns but following the Eastman Kodak "strategy," if we ignore their words and look carefully at their actions.

I see it all the time in my own business. I have heard Jim Rogers, former chairman and CEO of Duke Energy, speak very eloquently on the need to shift to renewable energy sources. In Duke's 2019 sustainability report, current CEO Lynn Good continues to spread the good word that the company is "becoming more efficient, more competitive, and more agile—while maintaining our commitment to sustainability."[56] But that message apparently never filtered down to the rank and file.

Duke happens to supply energy in the area where I live in Ohio. Duke is also about the best example there is of a power utility talking the talk but clearly walking … in the opposite direction. Whenever my company—and I am sure many others—tries to install a solar array in their service area, Duke seems to find all kinds of reasons to delay us from being able to connect to the grid. It could simply be bureaucracy of the highest order. But most everyone I know believes it is the utility industry's way of trying to stave off the clean energy revolution so the Dukes of the world can maintain their market dominance as long as possible.

We don't need to speculate about how Duke's public message contrasts with their actual corporate strategy, as there's ample evidence of Duke's intentions almost everywhere you look. In a 2019 report by the Environmental Working Group, coauthor Grant Smith, EWG's senior energy policy advisor, states it bluntly: "At a time when solar, wind, and energy storage costs are plummeting, Duke is seeking to slow the transition to renewable sources. It wants to retain control of power generation and protect its own power plant investments against competition and cheaper alternatives."[57]

One way Duke has earned that stinging assessment is by wielding their considerable political clout to derail state investments in clean energy and incentive programs for homeowner- and business-owned solar systems. One of their principal targets has been net metering—the means by which customers with solar installations can get credits from utilities for the solar power they send back to the electric grid. Duke wants their customers paying, not producing, so they do everything they can to thwart customers' ability to generate their own solar power and realize savings from those investments. For example, in its home state of North Carolina, Duke has more than doubled the flat monthly rate customers pay just for being hooked up to the grid.

They know that flat rate hikes discourage customers from investing in clean energy improvements because unavoidable fixed costs reduce savings and make up-front investments less attractive. And let's face it: net metering should be considered a subsidy to utilities rather than a payment for service rendered. Why utility investors—unlike shareholders in any other company—should have their investment protected is not explained. Because they always have, I guess.

In North Carolina, Duke has also lobbied successfully against legislation that would have made it easier to buy solar panels and stopped the extension of a renewable energy tax credit. But the company's efforts to fight the clean energy they pay lip service to in press releases are national in scope. According to federal campaign finance records, from 2014 to 2018 Duke spent more than $30 million, more than any other utility, to lobby the federal government. That included more than $726,000 in 2017 alone to push for the rollback of Obama-era clean air and climate regulations. An effort that was ultimately successful.[58]

Successful. I think we can agree that it's a relative term. Successful in slowing progress toward a healthier future? Successful in not innovating and reinventing themselves for the inevitable market realities, much like Kodak? It seems like years of operating as regulated monopolies have allowed utilities like Duke to stumble along, making steady profits now without developing the ability to compete moving forward. Their specialty is no longer producing reliable power at the lowest possible price. It is knowing what they can get away with under a complex set of rules they help to write. The utilities thrived in an ecosystem that demanded caution and compliance. Now that innovation and flexibility are necessary, they simply don't have the resources. Cost-effective distributed power technology takes the battle into unfamiliar territory for them. They have responded, for the most part, by trying to do what they do best—manipulate the rules to make their competitors' job as difficult as possible.

It's understandable for utilities to act defensively, as the problems with their business model are very real. In Europe, which is ahead of the United States in adopting renewables, utilities already are losing money. American utilities are headed down the same road. They know it. But unfortunately for them, they have no map to a better route. It's an old story in business. The utilities, like an organism, evolved to live in a particular environment. When the environment changes, the traits that made the organism thrive can suddenly become a hindrance.

I'd be afraid, too, if I were in their shoes.

POWER OUTAGE?

I'm quite sure there are voices inside Duke Energy and other utilities around this nation and throughout the world that are advocating for doing more than paying lip service to renewable energy. I'm sure there were those at Kodak who saw the digital camera revolution coming, those at AT&T who would have preferred to try obstructive lobbying and sticking to more of the same.

As Scott Smith, vice chairman of the US power and utilities team at Deloitte, has written in the *Wall Street Journal*, "Evolving technologies, customer preferences, and the competitive landscape are prompting business model changes for some power companies. New models often further enable a transition to clean energy; in some cases, power companies may also find new sources of revenue."[59]

The only way fossil fuel and utility companies can stop their death spiral is to stop acting like lobbying firms and start acting like businesses. Fire the lawyers whose job it is to protect the status quo, and hire scientists and engineers who can help the companies innovate their way out of their mess. The established players still have a huge advantage of incumbency and economies of scale in delivering power to the masses. Those that leverage this advantage can find new ways to become great.

Those that don't will soon be like Kodak, a name your children may not even recognize.

PART 3
THE SOLUTION IS CLEAN ENERGY

CHAPTER 8

We *Can* and *Must* Change

I n November of 2018, as dawn was just breaking over the Sierra Nevada mountains in Northern California, a firefighter radioed in to Cal Fire. He had spotted a small blaze near Poe Dam, a hydroelectric facility owned by the Pacific Gas and Electric Corporation (PG&E). "Probably ten acres, from what I can see," he reported. But within hours, high winds had acted like bellows, and the fire had exploded across the Plumas National Forest, engulfing thousands of acres.[60]

The resulting wildfire came to be known as the Camp Fire, the deadliest and most destructive wildfire in California history and the most expensive natural disaster in the world in 2018 in terms of insured losses. In the end, the blaze caused $16.5 billion of destruction, took eighty-five lives, and upended tens of thousands more. Individuals and families weren't the only casualties. The energy behemoth PG&E was forced into bankruptcy, becoming the S&P's first climate change casualty.

That same year wildfires raged in, of all places, the Arctic Circle. Sweden, the worst hit of all the Nordic nations, had to call for help from European partners like France and Spain, who were more accustomed to battling seasonal fires. With fires also sweeping across Norway, Finland, and Russia, there was no choice but to reach out to southern neighbors. The Swedes simply didn't have the planes and other wildfire infrastructure to battle the blazes. They were unprepared, because this had never happened.[61]

Meanwhile, in the southern hemisphere, uncontrollable fire swept through an estimated eleven million hectares in Australia, killing at least thirty-four people and destroying over six thousand buildings. The toll on Australian wildlife was even more stunning: an estimated 1.5 billion animals died in the blazes. Researchers are still tallying the damage and assessing the potential for recovery for many native plant and animal species.[62]

And as warmer temperatures put us in the path of fire, we find ourselves equally threatened by the retreat of ice. The Greenland ice sheet is now shedding about 267 billion metric tons of ice per year, while thawing permafrost is releasing three hundred to six hundred million tons of net carbon a year into the atmosphere.[63] The thaw of ice sheets at our planet's poles accelerates how quickly heat is absorbed, raises sea levels, amplifies warming with thawing permafrost, and disrupts ocean currents, feeding more extreme weather. Already, winter carbon emissions from the Arctic alone may be putting more carbon into the atmosphere than is taken up by all plant life on the planet each year.[64]

These are just a few of the more dramatic and recent stories from the frontlines of global warming. At this point it is no longer alarmist to say fossil fuels are poisoning our air, making our land uninhabitable, endangering our seas and drinking water, and destroying the biodiversity humans depend upon. The species headed toward extinction include "one-third of all reef-building corals, a third of all freshwater mollusks, a third of sharks and rays, a quarter of all mammals, a fifth of all reptiles, and a sixth of all birds."[65] Humans have left a destructive legacy, and it's getting worse. Climate change could mean the end for a sixth of all plant and animal species, or possibly more.[66]

For the sake of our health—and possibly even the survival of our species—we must change. I'm not saying the necessary change will be easy; if it were, we already would have done it. But the longer we wait, the more painful it will be, and the less likely we will be to avert truly devastating damage to our planet.

Of course, this is not a uniquely American problem, nor will the solution be driven by any one nation. We know China's economic growth has its demand for energy increasing exponentially. We also know India is not far behind. Over the next decade or two, the world's two most populous countries are going to spend trillions of dollars on energy infrastructure.[67] The more they invest in old, dirty energy sources, the harder it will be to switch to a more sustainable path.

The good news is that rising living standards in developing countries do not mean the world is doomed to follow the shortsighted path Western nations historically have taken. Instead of automatically turning to coal and oil as the cheapest options, developing countries can use solar, wind, and hydropower technology to avoid the dependence and economic instability that have plagued us, and China in particular is aggressively pursuing these options.

Yet it will be much easier for emerging nations, and for that matter the industrialized world, if the United States shakes off its resistance to change and leads the way. In fact, everyone on the planet would benefit, because pollution from dirty fuel is no longer a local problem. It is a global problem requiring worldwide action, and no country is better positioned to provide leadership and innovation focused on tackling this existential threat than the United States.

In 2014 Senator James Inhofe, chairman of the Environment and Public Works Committee, performed what became one of the most infamous climate change denial stunts in history. Inhofe placed a snowball in a plastic bag and brought it onto the floor of the US Senate. "We keep hearing that 2014 has been the warmest year on record," he exclaimed. "I ask the chair, you know what this is? It's a snowball, and that's just from outside here, so it's very, very cold out, very unseasonable." He proceeded to toss the snowball toward the chair, satisfied that he'd made his profound, drop-the-mic point.[68] Inhofe, in a key position of leadership on environmental policy, was theatrically confusing weather with climate. It was clear to most of those who heard him that the fact that he had had plenty to eat that night did not indicate that hunger had been eradicated.

This is why scientists today have learned to use the phrase "climate change" rather than "global warming." Warming is just the tip of the iceberg of what is actually happening to the planet. The weather outside one's window on any particular day is hardly evidence of what is to come. In the 1960s and 1970s, there were more record lows than highs in the United States. By the 1980s that had reversed. And in the first decade of the twenty-first century, there were twice as many record highs as lows.[69]

Another key objection to aggressive action to counter climate change runs generally along the lines of "sure, the climate might be changing, but are humans really responsible, or is this just nature's natural cycle, beyond our control?"

Well, multiple studies published in peer-reviewed scientific journals show that 97 percent or more of actively publishing climate scientists agree: climate-warming trends over the past century are largely the result of human activities.[70] More scientists disagree about whether or not smoking can lead to cancer. When 97 percent of experts worldwide reach the same conclusions, there's no room for controversy or even serious debate. Consensus crosses all lines— nationality, race, gender, political persuasion. Experts basing their position on research and demonstrable facts agree: climate change is happening, it is of the gravest concern, and human activity is the primary contributor. And yet, surveys by Yale and George Mason universities have found, only about 15 percent of Americans are aware that the expert climate consensus exceeds 90 percent.[71] If anyone doubts the effectiveness of the fossil fuel industry's disinformation efforts, that astonishing disparity between fact and perception should leave no doubt.

For those who question whether humans can do irreversible damage to the entire planet, let's try a thought experiment. How much harm can one person do? Well, it only takes one mishandled plastic bag ending up in the ocean to kill an unlucky seabird. Now think about how many bags you use every time you go to the grocery store and how many shopping trips you make every year. Think about how many other plastic items you throw away every day. Each one has the potential to harm a living being.

How much harm can a community do? There is no doubt a city can make itself unhealthy. We can all think of examples—from the eye-watering smog in Los Angeles and Beijing to Cleveland, Ohio, where the Cuyahoga River literally burst into flames.

What about a nation? To name a few examples, exposure to the herbicide Agent Orange, used by the United States during the Vietnam

War, was still causing birth defects in Vietnam thirty years later.[72] Soviet authorities evacuated one thousand square miles after the Chernobyl nuclear disaster. Three decades later, the population of that area—once more than one hundred thousand—is now just over one hundred.[73]

Extrapolate the environmental damage one person, one city, or one country can do to the seven billion people around the world—or the ten billion we're likely to have within fifty years. It is not arrogant to say humans can change the entire planet. It's arrogant to think we can't.

> **It is not arrogant to say humans can change the entire planet. It's arrogant to think we can't.**

Indeed, we already have. Fertilizer runoff from farms has caused hundreds of "dead zones" in the oceans, where the oxygen content is too low to support aquatic life.[74] Scientists fear that human activity is causing a loss of biodiversity the likes of which have only been seen five times in the past five hundred million years.[75]

The climate can change for a variety of reasons over hundreds or thousands of years. However, we have known about the heat-trapping properties of carbon dioxide since the nineteenth century.[76] It's clear that it is no coincidence that our current warming trend coincides with an unprecedented release of carbon dioxide.

The IPCC states, "The best estimate of the human-induced contribution to warming is similar to the observed warming over this period."[77] In other words, we did it. Do you think talk radio hosts and politicians from coal- and oil-producing states like Senator Inhofe are likely to know more about the topic than the world's leading climate scientists? While scientists cannot quantify with certainty how much warming we have contributed, there is no meaningful debate about the fact that human activity is responsible for rapid climate change that will radically alter the equilibrium of our planet.

For one thing, the trend is consistent. When scientists look at other effects on global temperatures, such as volcanic activity and solar cycles, they would expect some decades to be warmer and others to be cooler. Instead, we are now seeing each decade become warmer than the last. Some who seek to delink climate change from human behavior have hypothesized that the sun may be giving off more heat, yet if that were the case, we would expect more warming at the equator, which gets more sunlight, than at the poles. Instead, the opposite is happening. Also, the warming scientists see isn't evenly dispersed throughout the atmosphere. While the lower atmosphere is warming, the upper atmosphere is actually cooling. That points again to greenhouse gases trapping heat as the culprit.

As nice as added warmth might sound on a cold January day, the actual effects of adding more heat to the atmosphere are unpredictable and increasingly deadly. The temperature rise around the globe will not be even. Extreme heat waves, like the one that killed thousands in Europe in 2003, happen four times as frequently as they used to. If greenhouse gas emissions aren't brought under control, they could become sixty-two times as likely.[78]

On our current path, it is only a matter of time before droughts, heat waves, devastating hurricanes, and rising oceans make more and more of the planet unlivable at least some of the time. The point is that temperature changes that sound small could have catastrophic consequences. Scientists don't know how much temperature change would be tolerable or how much carbon we could emit before we reach a dangerous tipping point. Yet even as a certain amount of climate change is already inevitable, we continue to make the situation worse.

A very painful future might be very close. Already, Antarctica is losing ice the weight of Mount Everest every two years.[79] Because ice is white, it reflects sunlight and heat away from the planet. Replacing the white ice with blue water or brown land means less heat is reflected and more is absorbed. And though industrialized nations may be responsible for the vast majority of the carbon release that is warming the planet and melting the ice cap, the most dramatic impacts often fall upon people in remote areas who have had very little to do with causing it. The loss of sea ice in the Arctic means native Alaskan villages on the coast are eroding away.[80] And the rise in average sea levels—already expected to reach one to three feet by the end of this century[81]—are salinizing the fresh water in low-lying Pacific islands.[82]

As if the dramatic effects of retreating ice weren't bad enough, consider that the land itself is beginning to melt. Hundreds of billions of tons of carbon are trapped in the frozen Arctic tundra—more than is currently in the atmosphere.[83] As that starts to thaw, methane will gradually be released, and the global thermostat will be turned up even higher. To avoid the worst disasters, the IPCC suggests holding warming to about 2.0 degrees Celsius (about 3.6 degrees Fahrenheit) above preindustrial levels. To have a fifty-fifty chance of keeping the temperature at that level, we would have to keep carbon levels to about 450 ppm—and we already have moved more than halfway there.

To make things worse, there is a good chance the pace of warming will speed up. So far, much of the extra heat has gone into the oceans, which warm more slowly than air. The water eventually will warm to the point where it won't absorb as much heat, and the air temperature will rise instead.[84] Extreme heat has already killed a quarter of the coral in the Indian Ocean, along with the marine life that depended on it. Species that are able are moving toward the poles or into deeper waters to remain in the temperature range to which they've adapted and "advancing their breeding, hatching, budding, and migrating times."[85]

These are only some of the mechanisms by which the warming of Earth could create its own unstoppable momentum. Basically, the hotter we get, the more heat we trap. The warming oceans will evaporate more quickly, bringing more water vapor into the atmosphere. Water vapor is also a greenhouse gas, so it, too, will make the air warmer over time.

But, in the end, even if the consensus of almost every climate scientist in the world doesn't sway you, a move away from fossil fuels to natural energy harvesting is still the only reasonable play. The risks are asymmetrical. If we fight climate change and turn out to be wrong, we will still have invested in our companies' and country's future. We will have improved our economy, security, and environment and be prepared for the inevitable day when Earth's fossil fuel reserves are depleted.

If NASA announced there was a 1 percent chance of a city-size asteroid hitting Earth, wouldn't we want every government in the world looking for a way to save us from a civilization-threatening disaster? Wouldn't we want our scientists and engineers to throw their efforts at the problem? Of course we would. There would be no controversy, because there are no big asteroid companies making billions from space collisions.

Well, in the case of climate change, the catastrophic risk is far higher—in fact, it's a scientific certainty. If we ignore climate change, we will have more droughts, floods, heat waves, loss of species, disease, refugees, war, famine, and death. Therefore, our commitment to a clean energy future needs to become bipartisan and total, and it must start immediately.

We *can* and *must* change. The naysayers said we couldn't phase out sulfur emissions that caused acid rain or chlorofluorocarbons that damaged the ozone layer. We did, and the economy continued to thrive—thanks to the ingenuity of businesses with the right market incentives. Those who doubt the ability of the market to adjust to new realities such as placing a price on carbon are usually wrong. The sooner we stop listening to them, the better off we all will be.

CHAPTER 9

Energy Conservation and Efficiency– Use Less

"**A** penny saved is a penny earned." It's one of Benjamin Franklin's most famous adages, and none is truer. There are two ways to build wealth: by earning more and by spending less, and the principle is particularly germane to how we produce and "spend" energy. Up to this point, we've focused primarily on challenges with the production and distribution of energy—the problem of fossil fuel energy. But as we turn our attention to more encouraging topics—clean energy solutions and the promising future of renewable energy—we can't overlook a key segment of the energy economy that often goes unnoticed: the third-largest electricity resource in the

United States is, in fact, the penny saved: energy efficiency![86]

Even if you don't have solar panels on your roof, geothermal heating and cooling system, and low-E glass in every window of your home or business, you're profiting every day from the energy efficiency standards that began to be put in place in the 1970s after the oil embargo. It's estimated that the average American household saves $840 per year from energy efficiency standards built into their appliances, building codes, and utility licensing requirements. If those standards are made even higher and American families and businesses make a conscious effort to boost efficiency in the years to come, energy efficiency can become our largest electricity resource, providing as much as one-third of expected electricity generation needs. That kind of energy savings would avoid the need for almost five hundred power plants.[87]

Like getting your annual physical exam, improving your energy efficiency is one of those things that's easy to put off even when you know you need to do it. You have enough to worry about while running a household or a business, after all, and energy costs may seem like small potatoes compared to other strategic priorities. You can call your facilities manager next week with those questions you have about your building. In the meantime, money is flying out your leaky windows, burning in your inefficient furnaces, and paying fat dividends to your gas/electric utility.

Economist William Nordhaus calls this "energy-cost myopia." You can clearly see the short-term cost and hassle of making your building more efficient, but the long-term savings are blurry and off on the horizon. While understandable, this is an example of the kind of short-term thinking that has always plagued American businesses.[88] At least, it used to, but I'm happy to say that is changing. If you're a business leader reading this and considering rethinking your commitment to energy efficiency, you'll have some very good company.

Target, Apple Computer, Walmart, Starbucks, JP Morgan Chase, Samsung, General Motors, Berkshire Hathaway … virtually all of the country's business giants have integrated both efficiency and renewable energy into the way they do business. Clearly, these companies are looking beyond the next quarterly statement. They are investing in the future of their brands. They know sustainability is something consumers want and are increasingly demanding. And they know the elimi-nation of waste is the holy grail of long-term competitiveness. In the case of energy efficiency, they've realized that good and smart are the same thing.

> **The elimination of waste is the holy grail of long-term competitiveness.**

Conservation and energy efficiency make sense not just for *Fortune* 500 companies; this is also the right strategy for virtually any business's bottom line. With long-term financing, you might even increase your cash flow from day one. It is one of the safest invest-ments you can make, offering returns that would please any mutual fund manager. I know, because I have seen the benefits in my own company.

I should stop to point out that I've used the words *efficiency* and *conservation* interchangeably here, and while the two topics both fall under the general heading of using less, or "a penny saved," they are not exactly the same. Energy efficiency involves technology that doesn't generate electricity but reduces how much electricity is needed. LED lighting, variable-speed motors and automation controls as well as building insulation and double- or triple-glaze windows are some examples. They will never get a building to NZE by themselves, but these capital improvements are a major component of any serious move toward clean energy independence.

Energy conservation, on the other hand, requires no technology or investment—only thoughtful human behavior and good habits. Examples include turning off lights when they are not needed, turning thermostats down to sixty-eight degrees during the winter, and closing window blinds during the summer to reduce the air-conditioning load.

A commitment to both efficiency and conservation has an immediate impact. It allows you to move beyond thinking of your energy bills as fixed overhead. Like labor, materials, and other variable costs, energy is a cost you need to control. And energy efficiency and conservation are your quickest and most cost-efficient tickets to doing just that.

How much savings? Big savings. Studies show energy usage can often be cut by as much as 30 percent with a well-designed building retrofit. Most projects bring an internal rate of return greater than 15 percent.[89] Sure, in the best years you might do better in the stock market, but we all know that's a gamble, and these are solid, predictable investments in your company's future. Investments in efficiency are an excellent way to hedge against risk, including a risk that is often forgotten—rising energy prices.

To be sure, gains on that scale require more than changing a few light bulbs. The biggest savings will come from an examination that considers all of your building's systems as a whole rather than isolating each individually. That's what we did with our headquarters. For example, if we had focused only on getting more natural light into our building, we might have ended up with large, single-pane windows that raised our heating and cooling costs. If we had been concerned only with superinsulating our building envelope, we might have ended up with a box for a building, lit entirely by fluorescent bulbs. By thinking about our systems in a holistic way, we were able to multiply our savings.

The 15 to 30 percent or more of your current bill that you won't be sending to the utility company each month can be used instead for investments that grow your business. That's why experts say efficiency is not only an environmental policy but an economic policy that could create nearly two million jobs by 2050.[90] And you can keep that investment in your community instead of sending your money to an out-of-town utility or oil company.

The potential for economic stimulus is huge. There are more than eighty billion square feet of commercial building space in the United States, 30 percent of which is due for renovation anyway.[91] It's the perfect time to get the most bang for your buck. You can make your building newer, more valuable, and more efficient at the same time.

Because of energy-cost myopia, however, those eighty billion square feet are being retrofitted at a rate of barely 2 percent per year.[92] Getting the most for your energy dollar might seem like a no-brainer, and it is. But that doesn't mean everybody's doing it—yet. It might be as simple as business leaders not understanding what kind of opportunities they are missing. But there is still time to get a cost advantage over your competitors, or if they have already jumped ahead of you, now is your chance to catch up to the world's most successful brands.

Anyone who tells you environmentalism wants to deprive people of modern life is simply wrong.

To be sure, the potential of minimizing the energy we waste is low-hanging fruit and an exciting part of the new energy economy, but we cannot simply conserve our way to a healthier planet. The changes to our lifestyle would be too great. Nobody is signing up to give up their refrigerator or return to the days of handwashing clothes. Anyone who tells you environmentalism wants to deprive people of modern life is simply wrong. The fact is that with smart investments we can have the amenities we have grown used to in a way that won't compromise our grandchildren's ability to live even better. The efficiency revolution won't be as noticeable as the switch to renewable energy, but it will be every bit as important.

That's why efficiency is often called the "third fuel," after oil and natural gas. In fact, in the twenty-first century, we should be calling it the "first fuel," both for the priority we place on it and the impact it can make. The most common way of measuring the impact of conservation and efficiency as a "source" of energy is the energy intensity indicator, defined as energy use per real dollar of GDP. The dramatic effect that efficiency and conservation have on our energy economy is illustrated rather dramatically by these side-by-side charts:

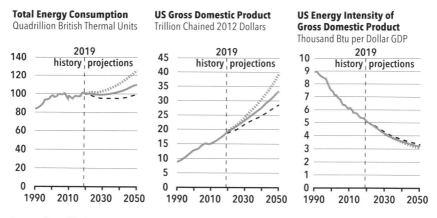

Source: Stacy MacIntyre, "EIA projects US energy intensity to continue declining, but at a slower rate," US Energy Information Administration's *Annual Energy Outlook 2020*, February 20, 2020, accessed June 14, 2020, https://www.eia.gov/todayinenergy/detail.php?id=42895.

The contrasting curves are dramatic. As our GDP rises, our energy use vis-à-vis output is plummeting. There's really no serious conversation to be had about where the future lies. Despite all attempts to slow its arrival, the future of energy-efficient buildings and renewable energy cannot be denied. It no longer really matters where we land on the political spectrum. It may seem ironic, but as we move into the century of electronic technology, we are using less electricity in relation to productivity, and one of the key explanations is that we are getting so much better at using energy efficiently. In fact, in the major industrialized countries, efficiency saves as much as the total energy use of the entire European Union. As the International Energy Agency puts it, "Energy efficiency savings in 11 IEA member countries were effectively displacing a continent's energy demand."[93]

Far from compromising our lifestyle in industrialized countries, efficiency has enhanced it. At the same time appliances such as refrigerators, dishwashers, and clothes washers have become more efficient, they have gotten less expensive and more reliable—with added features. Some other devices, including commercial air conditioners, are more expensive, but the added cost pays for itself several times over during its useful life.[94]

The United States and much of the world have largely done this without a sustained national effort. If we truly make energy efficiency a priority and extrapolate the potential gains across the globe, the effects will be revolutionary. As Amory Lovins, chairman and chief scientist of the Rocky Mountain Institute, has written, "Today, based on standard economic growth and decarbonization forecasts, cutting global energy intensity … by about 3–4 percent a year, versus the historic 1 percent, could more than offset net carbon growth and rapidly abate further climate damage."[95]

SO HOW DO WE DO IT?

For businesses, there are incentives for high-efficiency lighting, HVAC systems, information technology, and other equipment. It might seem strange that utilities would encourage people to buy less of their product. But if they can avoid building an expensive new power plant, they come out ahead. There truly are no losers with greater efficiency.

And so there also are no reasons to delay.

I've shared my own experience building out an energy-efficient headquarters for Melink Corporation. You might think that because we built our new headquarters from the ground up, it was easier for us to implement various energy-efficiency measures. But other than our southern orientation, every measure could have been implemented as a retrofit. In fact, for this reason I have implemented most of them in my own home too.

An important guide for us was the US Green Building Council's LEED certification program. This program provides a framework for helping building owners think about all the best practices in design, construction, operation, and maintenance for becoming more sustainable. Racking up LEED points wasn't our goal, but studying the standards gave us ideas on how to proceed and a benchmark for how well we were doing. When we added everything up, we had the first LEED Gold–certified office building in Ohio. The LEED standards offer dozens of ways to make our structures more sustainable, and they are flexible enough to be adapted to renovations as well as new buildings.

If you lease space rather than own, you may be unable to redesign your building. But that's no reason not to do what you can. You might also look into a "green lease" that provides for landlords and renters to share energy costs. That way, both owners and renters have an incentive to implement the most energy-efficient technologies available. The federal government's Energy Star rating system makes it easy to identify the most efficient electric appliances, so at least your plug loads are minimized. And conservation policies, such as water bottle-refilling stations, limiting hard copy printing, and recycling, are just as effective for lessees as they are for building owners. Every business can—indeed, must—create ways to become more efficient and thereby ensure their long-term success.

Whether you lead a start-up or a *Fortune* 100 corporation, planning for long-term success means maximizing efficiency in all areas: operations, logistics, customer service, you name it. Energy efficiency is a best business practice, but it is also more. Socially, ecologically, and morally it is the right thing to do. At the same time you are making your business more sustainable, you can make our planet more sustainable.

This is something no real leader should put off until tomorrow.

CHAPTER 10

Renewable Energy– Produce More

E nergy conservation can partially get us to the carbon-free economy that a healthy, prosperous future will require. Energy efficiency can take us another big step of the way. But to fully reach our goal, we will need to make the switch to 100 percent renewable energy production over the next ten to twenty years. A penny saved, yes, but a growing world population will require thousands and eventually millions of solar power– and wind power–generating facilities, among others, to maintain and improve our standard of living.

One aspect of the clean energy economy that I find most appealing is the level of grassroots, citizen-led participation it invites. It's not easy to drill for oil, dig for coal, or collect natural gas in your backyard. Fossil fuel exploration, extraction, processing, and distribution is, by its very nature, big business. Enormous multinational corporations provide the resources we can't produce ourselves, so we consume it at whatever price they set. There are no corner stores.

And as we transition to a renewable energy economy, we certainly need utilities and big energy companies to build out large-scale solar and wind farms, tidal turbines, and the like. But at the same time, one of the elegant features of clean energy is that every home and business in the world has the potential to become a mini power plant and in so doing see an increase in its property value. There are already tangible examples of this happening. Through a partnership between the real estate website Zillow and the solar energy rating company Sun Number, "solar potential" is now a factor in determining home values. Sun Number uses a variety of factors, such as the angle of the roof, the direction the home is facing, the solar radiation potential of the roof, and local electricity rates, to analyze an individual home's solar potential. They assign each home a score between 0 and 100, estimating how much homeowners might benefit if they install solar. A Zillow search shows my own home scores a Sun Number of 85.34, and my solar arrays are at work every day, raising the value of my home and qualifying me as a small power company. Feels like liberty.

> **One of the elegant features of clean energy is that every home and business in the world has the potential to become a mini power plant, and in so doing see an increase in its property value.**

Even if you're not personally sold on the idea of sustainability and "green energy" as a moral imperative, you'll arrive at the same place simply by analyzing the data and following the money. There's simply no question that this is the direction the world is headed.

Take a look at the chart below. It shows a dramatic split between investments in clean energy stocks and what the analysts euphemistically refer to as "natural resource" stocks. The trend lines and numbers paint a dramatic picture of where our energy future and the smart money are. Clean energy boasts stunning returns of more than 66 percent for the three-year period through February 18, 2020. Natural resources stocks—oil, coal, and natural gas—returned 11 percent over the same period.

Clean Energy Stocks Have Pulled Ahead of Natural Resources
Total Return, Three-Year Period through February 18

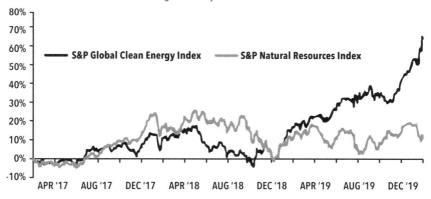

Source: S&P Dow Jones Indices, US Global Investors

Keep in mind, this data takes us through December 2019, before oil prices, for the first time in history, actually went negative in April of 2020. That means traders were actually paying people to take oil contracts off their hands. That's not a dip; that's a death knell. As *Bloomberg News* journalist Tom Randall puts it, "The reason solar power generation will increasingly dominate: it's a technology, not a fuel. As such, efficiency increases, and prices fall as time goes on. The price of Earth's limited fossil fuels tends to go the other direction."[96]

Randall's point is simple but incredibly important. Renewable energy is not merely an alternative to fossil fuels. This is not a question of simple preference—chicken or beef. Renewable energy is based on rapidly developing technologies that harvest sources of energy that never run out. Renewables are not limited sources of stored energy that we need to tear from the earth, fouling our environment as we do so. There really is only one source of energy on planet Earth, and that is the sun. When we put a log in the fireplace, turn on a gas stove, fill the tank of our cars, or fire up coal-burning power plants, we're tapping into the same energy source. Plants photosynthesize the sun's energy to grow, and burning wood or biomass releases that energy. Fossil fuels are concentrated plants and organic matter, storing solar energy absorbed millions of years ago.

So the question for our energy future is not whether we will use solar energy; human beings always have and always will. The question is whether we will use it in a form that takes thousands of millennia to build up, or a form that can be used in real time every day. Whether we use forms that pollute our air and change our climate, or forms that don't harm our health or force our children to adapt in unpredictable ways. Whether we use forms that become increasingly scarce and expensive, or forms that will be plentiful and increasingly affordable for the imaginable future.

A RENEWABLE ENERGY PRIMER

There are five major renewable fuels: solar, wind, geothermal, hydropower, and biomass—but every renewable energy source has its origin in solar power. The sun heats the ground, and that heat can be harvested for geothermal energy. The sun makes water evaporate, and that water comes back down in the form of rain, which eventually drains into rivers, whose kinetic energy can be captured by dams for hydroelectric power. The sun creates different temperatures in neighboring air masses, which cause wind that can be used to move turbines, which create electricity. And of course the sun can be used directly—either to heat and light our buildings directly or to be converted into electricity.

Each of these energy sources is valuable, and each has pros and cons. A sustainable energy future probably includes all five, but I'm going to focus primarily on the first three—solar, wind and geothermal—because those are the technologies individuals and companies can most readily harness to escape the status quo of volatile energy costs and reap the benefits of "going green."

There are two basic types of solar power: thermal and photovoltaic (PV). Our headquarters, for example, uses both. PV cells—the shiny panels you see on more and more rooftops—use light to dislodge electrons from a conductive material, creating an electric current. Solar thermal power concentrates heat from the sun to warm a liquid. We use it to heat water for our building. The technology also can be used to create steam that drives a turbine to create electricity.

The best feature of solar PV cells is that they have no moving parts, so they require virtually no maintenance. If you can put them on your roof, they require no additional land. The downside, of course, is that they only work when the sun is out, so PV provides maximum benefit in areas with abundant and consistent sunshine. It's also ideal for schools and businesses that only operate during the day. That means that solar, for the time being, needs to be combined with other energy sources to assure consistent power under all conditions. The future of solar, however, will be combined with storage solutions that can provide dependable solar power even when the sun doesn't shine.

Wind is a nice complement to solar energy, as it can work any time of day, though only when the wind is blowing—again, storage solutions will even this out. Our building draws power from the utility grid to supplement our production on a cloudy day or a still night.

But over the course of a year, we generate more electricity than we use.

Geothermal heat pumps use the constant temperature underground for heating and cooling. Basically, the ground provides heat in the winter and absorbs heat in the summer. The system requires some electricity to run but returns three to five times as much energy as it requires in the form of heat transferred to or from the building, depending on the season. Geothermal energy can also be used to generate electricity. Volcanic Iceland depends heavily on geothermal power, but that is typically for utility-scale generation. We don't have a volcano outside of our facility, but our geothermal heat pumps kick on before the sun comes up, using electricity from a bank of old truck batteries that store the excess power we generate during daylight hours. Again, the beauty of renewable energy sources is how they scale down, from a volcano producing energy for urban utilities right down to a bank of truck batteries powering a single business in Milford, Ohio. Self-sufficiency and freedom.

But can renewable energy power the world? Unquestionably yes, many times over. Earth absorbs enough energy from the sun in about an hour to power the world for a year.[97] Yet, as of 2018, solar and wind accounted for less than 4 percent of all energy used in the United States. Fossil fuels fueled about 80 percent of the nation's energy demand.[98] That percentage is dropping, but not fast enough to address the catastrophic consequences of a warming planet. Impediments, both real and manufactured, are slowing our progress, and for many families and businesses, a clean energy revolution is still waiting for leaders like you.

For renewable energy sources to reach the scale we need, they have to be seen as mainstream fuels. We are, after all, practical people. If renewable energy can't compete with the fossil fuels that we've long depended on for our growth and prosperity, then, for a majority of people, this is just an academic discussion. And that does seem to be the thinking, particularly in conservative circles, where renewable energy has been cast as a liberal, impractical fantasy.

But here is the astonishing truth. We are already there. The cost of renewable energy has plummeted to the point where almost every source of green energy is competitive with oil, coal, and gas-fired power plants.[ii] It's referred to as grid parity, and the data trends show that soon there will be no contest, with clean energy easily outdistancing fossil fuels on a simple cost basis, regardless of all the other obvious benefits to our environment.

For proof the new energy economy will work, we won't have to wait long. All we have to do is look to Europe, where several countries are far ahead of us in making the transition. Germany, for example, gets nearly half of its electricity from renewable sources and is on track to increase that to 65% by 2030.[99]

And the aggressive push toward a renewable energy economy hasn't hurt the country's economic competitiveness. As I write this, global economic statistics are dramatically skewed by the effects of the COVID-19 pandemic that have slowed economies worldwide, but 2019 pre–COVID-19 figures show Germany's unemployment rate stood at 3.1 percent,[100] even lower than the historic US figure of 3.6 percent[101] and one of the lowest in the Eurozone. Germany also enjoys Europe's largest economy and highest foreign trade surplus.

---

ii   As of May 2019, hydroelectric power was the cheapest source of renewable energy, at an average of $0.05 per kilowatt-hour (kWh). The average cost of onshore wind, solar photovoltaic (PV), biomass, or geothermal energy plants is now usually below $0.10/kWh. Not far behind that is offshore wind, which costs close to $0.13/kWh. The entire portfolio of green energy sources has reached grid parity with fossil fuels such as oil and gas, which typically range from $0.05/kWh to over $0.15/kWh. Dominic Dudley, "Renewable Energy Costs Take Another Tumble, Making Fossil Fuels Look More Expensive Than Ever," *Forbes*, May 29, 2019, accessed June 18, 2020, https://www.forbes.com/sites/dominicdudley/2019/05/29/renewable-energy-costs-tumble/#3207530de8ce.

One neighbor with similarly low unemployment is Denmark, which is even more of a standout when it comes to renewable energy. More than twenty years ago, Denmark built the world's first offshore wind farm.[102] By 2019 the Scandinavian country had passed the 50 percent mark for energy produced by renewable sources. Remarkably, half of renewable electricity production is solar, the other half exclusively wind.[103] In fact, when the wind is really whipping, Denmark's turbines can produce more power than the country uses—allowing it to export clean energy to Sweden and Norway. The Danes plan to be fossil fuel–free by 2050,[104] and thanks to that commitment, Denmark is now a leading exporter of wind turbines. Its wind industry employs more than twenty-eight thousand people[105] and accounts for more than 6 percent of the country's exports.[106]

As these examples show, clean energy–powered economies work, and those countries that were the first to embrace renewables are never going to turn back. But with the technology improving and getting cheaper, you might ask, "What's the hurry? The sun will still be there in five or ten years. Why can't I put up solar panels then for less than they would cost now? In fact, why should I put them up at all? If renewables really are economical, wouldn't the utilities install them and phase out their polluting power plants?"

I have two answers to those questions. One is that the greatest opportunities are likely to go to those who shape this emerging industry rather than let themselves be shaped by it. If American businesses wait to embrace renewable energy as consumers, then we lose the chance to make our country a leader on the production side. Other countries will be eager to take over that role. As of 2018 China was the biggest producer of solar technology by far, producing an eye-popping 73 percent of the world's PV solar cells. In that same year, the United States produced 1 percent. Room for improvement.[107]

The clean energy opportunity is still young enough that we could become an exporter of solar panels to China rather than an importer. Yes, government policies could help, but no real leader should wait for the government to show the way when the way forward is already clear. How long would we have to wait for our federal government to set any kind of clear, consistent policy? Probably far too long, painful experience tells us.

And even if the United States were able to break free of the fossil fuel lobby, to make a strong national commitment to renewable energy as our allies and adversaries alike have done, any sensible policy would rely on the private sector to provide the ingenuity to bring game-changing technologies to the market and at a scale that would make them affordable. American companies that invest in renewable energy for themselves will also be investing in what could be a major growth industry for their country.

My second answer is that we have the opportunity to change the fuels we use for energy along with the entire system for delivering it. Our current electric grid—powered by large plants connected by hundreds of thousands of miles of lines—is based on a twentieth-century industrial model. The interconnected nature of this vast, aging infrastructure leaves us vulnerable to outages that can shut down entire regions of the country, as in the 2003 blackout that left fifty million people without power in the Northeast.

That started with a software glitch, but it showed how damaging a failure could be at any number of choke points in the system. Whether it's by accident, natural causes, or a terrorist attack, major damage to a power plant or substation could cause a cascade of problems for hundreds of miles around. Weather-related outages alone do at least $18 billion worth of damage every year in the United States.[108]

And despite claims that the intermittency of some renewables could lead to even more instability in the grid, there are few signs of that actually happening—quite the opposite, in fact. Germany's power grid stability and security of supply has been excellent over recent years despite their huge expansion of intermittent green electricity production. Consider that in 2017, average power outages per German consumer amounted to just over fifteen minutes.[109] During that same year, electric power for US customers was interrupted for an average of 7.8 hours (470 minutes), or roughly thirty times longer than for German utility customers.[110]

As we scale up our use of renewables, the issue of intermittency will be effectively removed by the fact that we have a nationwide delivery system of fossil fuel energy in place. Those who argue that solar and wind power are not dependable enough to rely on all the time, that they will never be consistently capable of meeting the world's needs, can take comfort in the fact that natural gas and other fossil fuels can fill the gaps when the sun isn't shining and the wind isn't blowing. As we build out more and more renewable capacity, as storage technology develops, as costs continue to come down, as communities and utilities transition to smart metering microgrids, our petroleum and gas industries will continue to even out supply even as the percentage of energy they supply drops each year. This is a revolution, but there's no reason for it to be a bloody one.

That network of microgrids will be a much safer, more reliable distribution system that follows the twenty-first century model of the internet. There is no centralized internet plant that could knock out service to an entire city if it goes down. Instead, there are millions of servers. And even though they are connected, when one goes down, the problem is localized. That redundancy is the secret to the internet's resiliency.

Our electrical infrastructure can be that resilient, too, when power generation is distributed rather than centralized. When homes, factories, and office buildings are all generating their own power, power plants and substations will no longer be pillars holding up our entire economy. They might serve as backups to distributed solar and wind energy—although, as power storage technology improves, even that will be less important. But no power outage would be able to bring society to an instant halt.

When there is an outage in the area around our company's headquarters now, we lose power too, even if we're generating more than enough for ourselves. Because our surplus power is sold back to the grid, we are required to shut down when there's a problem so utility workers can safely repair it. But in the near future, our business park will have its own microgrid that connects and disconnects from the local utility as needed. True autonomy: this is clearly the future.

Renewable energy deployment, as we've discussed in some detail, is clearly going global—a source of hope for humanity and the future of our planet. But America is behind, not because it's not up to the challenge but for all the reasons we've already examined—principally because dirty energy has spent a lot of money kneecapping progress toward clean energy through disinformation campaigns and lobbying. But the fact is that opportunities to lead this energy boom are, as Rockefeller discovered at the outset of the oil century, particularly grand for Americans.

One thing the United States has plenty of is land. Solar and wind energy give us the chance to put sparsely populated land to use in a way that does not tear up the landscape or destroy wildlife habitat. Think of Germany's leadership in solar despite the fact that Germans vacation in Spain and Argentina; it's not a particularly sun-drenched nation. By contrast, our deserts across the Southwest are perfect locations for solar arrays. Wind turbines can be placed across our vast plains, or offshore on our two ocean coasts or the Great Lakes. We are as rich in resources for the clean energy revolution as we were for the Industrial Revolution.

And we have another resource even more precious that is made for this challenge. The history of the American people has always been one of innovation, creativity, and pioneering exploration. Elon Musk came to America to build a fortune, lead the electric car revolution, and aim for Mars because he knew this is the one place on Earth where dreams that big could come true. Our universities are incubators for research and start-up companies like no others in the world. This is the world's greatest economic engine. Writing the next chapter has always been what America does. It wouldn't just be surprising if we didn't get in the clean energy game to win; it would be a relinquishment of our birthright. The amount of energy available to us is infinite. Yet out of inertia and lack of imagination, we rely on sources that are difficult to access, harmful to use, and doomed to become increasingly scarce. This isn't the way forward for any nation, certainly not for the United States of America.

The technology to make it happen exists today, and it will only get better. Wind turbines and solar PV panels will become more efficient. Batteries for power storage will get cheaper. There will be breakthroughs in geothermal heating and cooling, hydrogen cells, storage. We should be the nation to lead this wave of innovation; the opportunity is enormous. Just consider the history of solar photovoltaics, which originated in the 1950s but were initially too expensive for commercial use. In 1977 the price per watt of silicon solar cells was $76.67. Over time, improved efficiency drove down prices, and early adopters began to invest. Demand dropped prices even further, starting a feedback loop. By 2016, solar cell prices had fallen to $0.26 per watt, a drop in cost of 99.6 percent in thirty-nine years. It's an opportunity we should be excited about, particularly as business leaders.

Already, smart companies around the world are realizing this and building a better future for themselves and their stakeholders. Huge advantages are there for companies that make energy independence a priority, and fortunes are there to be made for innovators in tech and finance who solve the remaining challenges. There is no reason to let your company lag.

All it takes to get ahead of the game is the vision to see a better future, the courage to push further than others have gone, and the wisdom to ignore those who say it can't be done.

CHAPTER 11

Free Markets and the New Gold Rush

A s we have seen, clean energy is a catchall for technologies that must help us kick the fossil fuel habit. Innovators are making these technologies more cost effective and accessible all the time.

The innovators I'm talking about are not only scientists and engineers. Equally important are the entrepreneurs and financiers who turn these technologies into mainstream solutions and profitable businesses. From Main Street to Wall Street, a growing number of business leaders are jumping into the clean energy space to make their fortunes.

This is happening because the market for these technologies is expanding exponentially. Customers love the no-risk, predictable returns of the sun rising every morning and the wind blowing every night. And new financial models from power purchase agreements (PPAs) to sale leasebacks are increasingly allowing these customers to buy nonpolluting electricity at or even below retail rates without the capital outlay for installing a solar or wind farm.

This sort of business innovation gives those of us who are not R&D scientists a way to help accelerate the clean energy economy. We don't have to become owners of these technologies … we just have to be willing buyers of the clean energy they generate. It's a win-win for both producers and consumers! But this has only been made possible with the help of various private sector and government incentives. Let me provide a quick overview of them … and then we can talk about politics and free markets.

FINANCING TOOLS

The exact mix of incentives varies by state and locality as well as by utility, and it's changing all the time. Renewable energy developers and installers are usually familiar with this landscape and able to fill you in on the specifics.

Net Metering

Net metering is the reason I look forward to getting our company's electric bill each month. When our solar panels and wind turbine produce more electricity than our building uses, our electric meter essentially spins backward—and our local utility pays us the wholesale price for the electrons we send back to the grid in the form of a credit.

Most states have some provision for net metering, and we're fortunate Ohio's law is one of the strongest in the country. Unfortunately, these laws have come under attack in some states. Utilities argue they have to recover the cost of maintaining the power grid, but in Ohio that is reflected in the fact that we import power at retail rates and export power at wholesale rates.

It's true—we will have to make some provision to pay for upkeep of the grid if enough people stop buying from the utilities. Similarly, we will need some method other than the gasoline tax to help pay for our roads when electric vehicles become the norm. But I look forward to that day, and net metering will help us get there. We're far from reaching the point where utilities are being starved of the money they need to maintain the grid. Their lobbying against net metering is one more instance of attempting to protect a business model that is increasingly out of touch with customer needs.

Time-of-Use Pricing

Perhaps the most decrepit business model used by utilities is charging the same price for a kilowatt-hour no matter when it is consumed. This is because the price tag utilities charge each other for power constantly changes based on supply and demand. So it's crazy that the customers who create demand have no idea what the real price is and have no reason to care. Since rates are always the same, customers have no incentive to conserve power when demand is highest—typically on hot summer days. As a consequence, the utilities need to have more generating capacity available than they usually use, raising costs for everybody.

The technology exists to change this, and some utilities are starting to use it. With time-of-use (sometimes called time-of-day) pricing, customers pay more only when power actually costs more. We've become accustomed to this kind of "surge pricing" with rideshare companies like Uber and Lyft. But for the time being, these programs are typically voluntary with power consumption, and the details vary depending on the utility. Not only do these plans make business sense;

they are also a boon for the environment. They promote conservation and efficiency by discouraging energy use at peak times, and they can provide an additional incentive to install solar power.

Because demand for electricity peaks in the middle of hot summer days, the price of power is highest exactly when solar panels are producing the most energy. If you are an energy-savvy businessperson, you should jump at the chance to sell some of the power that your competitors need to buy—and do it at premium prices. The more people embrace time-of-use pricing, the more the incentive utilities have to build out their own solar production capacity. And the more utilities invest in solar capacity, the less coal and gas they'll need to burn in the summer. That means less of a spike in prices at times of peak demand. Up to this point, most utilities have dragged their feet when it comes to embracing renewable energy. That gives many of us an opportunity to take advantage of their complacency.

Third-Party Financing

If the initial investment is keeping your company from installing renewable energy sources, it shouldn't. New business models allow you to take advantage of solar power without the expense of buying and installing panels. Two methods of doing this are leasing and PPAs.

With a PPA, a solar contractor will install a solar power system at little or no cost to the site owner. In exchange, the customer agrees to buy the power generated, usually for less than the local utility would charge. Terms of such agreements can vary. The length typically runs from ten to twenty-five years, during which the developer is responsible for maintenance of the system. Rates can be fixed for the length of the contract or rise at a scheduled rate. Either way, the customer will have a predictable price for electricity and will be able to take advantage of net metering.

The savings can start right away. And with little or no initial outlay, concerns about access to capital or length of payback period are moot. The customer won't own the system, but the agreement can be transferable. That can raise the value of the property. And while any tax credits for installing the system typically go to the contractor, the incentives still lower the price of power without creating any extra paperwork for the customer.

Some states put a hurdle in front of PPAs by regulating solar developers when they sell power. A way around this is by leasing a solar array, which has the same advantages: little or no up-front investment, predictable costs, and increased property value. The difference is that instead of paying the developer the amount of an electric bill, the customer makes a lease payment.

By no means am I trying to talk you out of buying a solar power array. When you consider the cost certainty, the useful life of the equipment, and the minimal maintenance, it is an excellent investment. Even taking out a bank loan to make that investment may well make sense. But if access to capital is an issue, there are ways around it that still allow for the top-line and bottom-line advantages of using renewable energy. That's why, in many places, most residential solar installations use third-party financing. There's no reason businesses can't take advantage of this option as well.

Renewable Energy Credits

One advantage of owning a renewable energy source is that electricity isn't the only benefit you get out of it. For every thousand kilowatt-hours of renewable energy you produce, you can receive what is called a renewable energy certificate (REC). A REC is exactly what it sounds like—certification that renewable energy was produced. RECs are certified through a third-party registration service. The most common standard is Green-e, administered by the nonprofit Center for Resource Solutions.

What can you do with these certificates? Sell them through a broker, further helping defray the cost of your investment. Companies will buy RECs for two reasons: to show them off or because they must. Utilities enter the market for RECs when they are required to include a certain amount of renewable power in their energy portfolios.

Most states have some sort of RPS. If the utilities don't produce enough renewable power with their own generating capacity, they can pay someone else to do it. That someone could be you, and that's where RECs come in. The certificates allow the utilities to claim credit for someone else's investment. In addition, many large companies buy RECs to support renewable energy and prove their sustainability commitment.

The price of RECs varies by state. REC value increases as more aggressive renewable portfolio standards drive up demand. Sometimes the method of generation affects the price. In some states, for example, utilities must include a certain amount of solar energy as part of their portfolio so solar RECs can sell at a premium.

PACE

Property assessed clean energy (PACE) financing can be another way to pay for renewable energy with little or no money down. Under this program, local government entities issue bonds, then loan the money to consumers and businesses to pay for renewable energy projects. Borrowers then repay the loan through a rider on their property tax bills. Most states have enacted laws making some provision for PACE financing. But even in those states, many local governments have not yet embraced the program. For small- and medium-sized businesses that have it available, PACE can be a worthwhile option to investigate.

In addition to extending the payment term over twenty to thirty years, which reduces the monthly cost, PACE is a way to finance energy projects off your balance sheet. In other words, the loan against your buildings does not affect your capacity for other debt, such as the line of credit you might rely on for working capital.

Investment Tax Credit

Perhaps the most high-profile and controversial incentive for renewable energy is the federal investment tax credit, or ITC. This provision, originally signed into law by President George W. Bush in 2005, gave companies a tax credit of 30 percent of the cost of installing solar or small wind-energy systems and 10 percent for geothermal and combined heat and power systems. Because of the popularity and effectiveness of these ITCs, Congress has extended them several times. While the 30 percent incentive carried through 2019, it dropped to 26 percent in 2020, will drop to 22 percent in 2021, and if Congress takes no action, there will be no incentives for residential solar starting in 2022. Commercial solar will continue to receive a 10 percent ITC.[111]

Obviously, a 30 percent tax credit has been a real boost for the solar PV industry, and businesses, schools, and homeowners across the country have benefited as well. But the real game changer is that the gigawatts of solar capacity installed, as a result, have helped reduce the cost of solar power over 80 percent in the last ten years. And this will continue to create economic and job growth for decades to come, even after the tax credit winds down.

The word *subsidy* gets many people agitated. When those subsidies clash with the interests of powerful, entrenched industries like utilities, coal, and oil, you have a real political football. One of my managers wrote an op-ed column in the local newspaper describing the benefits of his personal investments in solar power and an electric vehicle. For his trouble, some online commentators called him a freeloader, looking to the taxpayer for a handout.

Allow me to say a few words in favor of the ITC. First, it has always been a temporary measure intended to help grow a new industry that requires scale to be competitive with the monopolistic utilities. It has worked quite well. Thanks to economies of scale the tax credit has helped create, solar power is now competitive with traditional fossil fuel power in many places—even without a subsidy. And the downward-sloping cost curve will only continue to get better over time.

Second, every new energy source has received government support. The government has provided pipelines for oil and gas, and dams for hydropower. It has surveyed coal resources and funded the research that made nuclear energy possible. To expect renewable fuels to compete with established competitors who have relied on government support dating back to the early days of our republic is more than a little unfair. Indeed, the fossil fuel companies still get billions in subsidies despite being some of the largest and wealthiest multinational corporations in the world.

Even more important is the hidden subsidy the fossil fuel industry gets from being allowed to pollute our air and change our climate. You and I aren't allowed to throw our garbage into our neighbors' yards. And I know that my business isn't allowed to dump waste into the Little Miami River, even though that would be cheaper than hauling it to a landfill. The reason is obvious. Why should my neighbors have to deal with the mess I make? Yet that is exactly what we allow the fossil fuel industry to do. The costs of treating asthma exacerbated by diesel trucks and coal-fired power plants are costs private companies have foisted onto the public. Everything spent to move cities flooded by rising sea levels is a subsidy for fossil fuel use.

As a fiscal conservative and believer in free markets, I think that a tax credit for investments that clean our air instead of soiling it, protect our climate instead of changing it, and help improve industry competition instead of preventing it is more than compatible with a responsible use of government. Ensuring economic growth and a more affordable, less dangerous power supply—hardly a government handout.

If the ITC does not fully level the playing field between fossil fuels and renewable energy, it helps narrow the gap. It doesn't give clean energy an unfair advantage; it helps mitigate the advantage fossil fuels have enjoyed for many decades. All of that is to say that investing in renewable energy is not freeloading. It's the only way to build an economy in which customers truly pay the costs of the goods and services they use instead of passing them off to society at large.

> **Ensuring economic growth and a more affordable, less dangerous power supply–hardly a government handout.**

FREE MARKETS

The beauty of capitalism is that it gives people an incentive to do things that are socially useful—work, invest, innovate, delay gratification. Cheap fossil fuels have exactly the opposite effect. They give people an incentive to be wasteful, sacrifice the climate of the future for comfort today, and stop looking for a better way to do things.

Markets are magnificent machines. And when they are put to proper use, markets make life better in innumerable ways. But the fossil fuel economy has pointed the machinery of capitalism in the wrong direction. It has built personal fortunes for an elite group of executives and investors by impoverishing the future for everyone else.

It has privatized profits while socializing the security risks it creates. It imposes costs on people who receive no benefits. None of this is part of the Economics 101 model of how markets are supposed to work.

In a perfect world, the short-term benefits of switching to renewable energy would be as great as the long-term benefits, and the payoff for the customer would be as great as the payoff for the world. We wouldn't need long-term thinking or sacrifice; the right thing to do would also be the easy thing to do. And, therefore, it would also be the obvious thing to do—simply let the free market guide us to the clean energy future.

Unfortunately, as compelling as the case is for renewable energy, it requires constant vigilance to counter all the fossil fuel industry's disinformation. Even those of us who believe that the free market is central to American prosperity and progress understand that when markets are used in socially harmful ways, it is only proper for government to step in and nudge them to use their not so "invisible hand" for good.

And the most effective nudge in this case would be putting a price on carbon. Any economist will tell you that we'll get less of something if we tax it. For example, when it became clear that the tobacco industry had been hiding the addictive and cancerous effects of nicotine, even engineering tobacco to be more addictive, the federal government both sued for settlements to be redirected to healthcare and imposed steep taxes on tobacco products going forward. In the years that followed, cigarette sales plummeted, particularly among younger people, which led to equally dramatic drops in smoking-related ailments. Public information campaigns, support for smoking cessation, and restrictions on smoking in public places were all contributing factors, but a wide body of research has established that applying excise taxes on tobacco products has been the most effective policy instrument to discourage smoking.[112]

That is why I support putting a price on the harm fossil fuels do to our health, our security, and our environment. I support this not because I don't believe in markets, but because I do.

Having been in business for over thirty years, I know the private sector is a powerful agent for change. A carbon tax or a cap-and-trade system would give the market the long-term signal it needs. Think about what we are taxing instead. Work. Hiring. Investment. Exactly what we want more of in our economy. Shifting our tax burden away from income and payroll and toward carbon would not shackle the free market; it would unleash it. Instead of complicating the tax code with more credits and deductions, we would simplify it.

As a bonus, incentivizing families, property owners, and businesses to make the leap to renewable energy and electric vehicles would clear one of the last remaining obstacles to creating the clean energy revolution. No longer would there be any question of whether investing in renewable energy makes financial sense. Like that smoker who had been meaning to get around to quitting eventually—when prices rose and the government focused on promoting public health, that decision got a whole lot easier. It's time for us to get behind a similar push to pivot toward renewable sources of energy.

Payback periods would be shortened, and the race to improve renewable technologies would heat up. The same goes for conservation and efficiency efforts. A carbon tax would send a strong signal that there is no going back. No fracking boom or OPEC decision would bail fossil fuels out of their natural and inevitable decline.

In short, a carbon tax would be the quickest way to unleash the clean energy revolution. It makes so much sense that some companies are already planning for the day governments get their act together and start putting a price on carbon.

In fact, a number of the world's leading oil and gas multinationals, including ExxonMobil, British Petroleum, and Royal Dutch Shell, recently joined representatives from seventy-five *Fortune* 500 companies on Capitol Hill in an effort to lobby Congress for a carbon tax. It would have been an unthinkable scenario just a year earlier, but CEOs from every sector, from retail to agribusiness, were there because they were already taking hits to their bottom lines from the effects of climate change and had concluded that a carbon price was the only market-driven route to solving the problem of US emissions. Fossil fuel execs, who had historically opposed ideas like a per-barrel fee or a carbon tax, surprised many by aligning themselves with that position. They had come to terms with the fact that any system that allowed for the continued use of fossil fuels was better than even tougher restrictions that would put the United States on a path moving away from oil and gas altogether.[113]

If ExxonMobil, British Petroleum, and the rest of these oil and gas giants want a price on carbon, just like the clean energy industry … and the rest of the world, what are we waiting for?

THE NEW GOLD RUSH

While I'm hopeful that a carbon tax gains traction in statehouses and in Washington, business leaders and investors need not wait to forge boldly on. The more we take the initiative to move the world into the clean energy economy, the more we will be successful, and the more we will change the world for the better. Climate change, the need for more efficient storage and transmission technologies, rebuilding our grids and infrastructure … these are all problems that need to be solved. But entrepreneurs know that another word for *problem* is *opportunity*.

It's no secret there is money to be made in energy. The most profitable company in the world, by a healthy margin, is Saudi Aramco.

The top-twenty list for corporate profitability in 2019 also includes Royal Dutch Shell, Gazprom, and ExxonMobil. But if you're looking for where the greatest growth and opportunities lie moving forward, consider this *Financial Times* analysis of investment trends in 2019: "Investors who bet on a shift from fossil fuels to clean energy are being richly rewarded as solar and wind stocks outperform oil and gas shares by a widening margin this year. The iShares Clean Energy exchange-traded fund has risen by 32 per cent so far this year, streaking far ahead of the oil-dominated Vanguard Energy ETF, which has risen by only 1 per cent."[114]

What this means is that the move from fossil fuels to a clean energy economy has completed the journey from long shot to sure thing. Even the transition from the clean-fuels Obama administration to the dirty-fuels Trump administration, which has rolled back environmental standards and promoted coal, has done little to alter the calculus for investors and corporate planners. Investments in renewable energy continue to set records in the United States and in markets around the world. Every component of the renewable energy sector has been benefiting from a sharp reduction in the cost of wind and solar in recent years, which has made them cheaper than coal and natural gas in many markets. The cost of solar has fallen 85 percent since 2010, while wind power has dropped about 50 percent.[115] Those advances in technology, scale, and efficiency are so rapid and dramatic that they simply can't be ignored.

As I mentioned earlier, one of the factors that has kept fossil fuels indispensable in the past has been the fact that renewable energy sources like solar and wind are intermittent, dependent on time of day and weather conditions. But this concern is quickly being laid to rest with steady progress being made on battery storage technologies. As the international consulting firm Deloitte notes in its 2020 report

Renewable Energy Industry Outlook, "The greatest [price] decline was in lithium-ion battery storage, which fell 35 percent [in 2019]. This steady decline of prices for battery storage has begun to add value to renewables, making intermittent wind and solar increasingly competitive with traditional, 'dispatchable' energy sources."[116]

It is perhaps understandable that the world's major oil and gas companies have talked a big game about investing in renewable energy but have been slow to commit wholeheartedly to a clean energy future. After all, they are the descendants of Rockefeller, and fossil fuels have been the world's most consistently productive cash cow for over a century. But they would be wise to heed the cautionary tale of a nationwide chain of movie rental stores called Blockbuster that once dominated the distribution of films for at-home viewing. In 2000, when online streaming was becoming increasingly viable as internet-ready screens and internet bandwidth developed, Blockbuster turned down an offer to buy a fledgling streaming media service called Netflix for $50 million. Ten years later, Blockbuster closed its doors for good, and as I write this in 2020, Netflix is valued at over $215 billion. The technology curve, customer demand, and distribution costs were clear, but Blockbuster had enjoyed a near monopoly in a very profitable space that it didn't want to relinquish. This niche company's entire history lasted twenty-five years. Imagine how an industry unimaginably more profitable, international, and entrenched, an industry that powers and props up national economies around the world, an industry more than a century old, is tempted to fight and kill off rather than embrace clean energy. But they will do so at the risk of their own demise.

The future is clean energy. There is no doubt that the countries and companies that seize this once-in-a-century megaopportunity will be the ones that thrive for decades to come. And the United States shouldn't just be taking part in this revolution; we should be leading it. My call to America, particularly to conservative America, is that we must not deny the data, the clearheaded benefits, and the evidence of our eyes. This is not a partisan issue. Only the details of how we move forward, how we balance government regulation and business incentives, should be. This is where the conservative and liberal debate should take place. But we must all move forward to the same goal together, as we have done when fighting for freedom, when reaching for the stars, when we have been at our best.

> My call to America, particularly to conservative America, is that we must not deny the data, the clearheaded benefits, and the evidence of our eyes. This is not a partisan issue.

I am reminded of a story journalist Thomas Friedman tells in his book *Hot, Flat and Crowded*. While addressing an audience in China, Friedman talks about how he is often told the United States and other Western countries grew rich by burning coal and oil, and now it is China's turn.

Friedman replies, "Please, take your time, grow as dirty as you like for as long as you like. Take your time! Please! Because I think my country needs only five years to invent all the clean power and energy efficiency tools that you, China, will need to avoid choking on pollution, and then we are going to come over and sell them all to you."[117]

Friedman's audience knew he was joking. But there was a lot of truth in his words. The fact is, the race isn't only to lead in developing and deploying green energy technologies; the race is to see which countries will be first to recognize the opportunity, get their citizens on board, and make the transition a national priority. Friedman's story suggests that he has great faith in America's capacity to rise to the occasion, but he also acknowledges that the United States has arrived a little late to the party.

In many ways, Friedman might be pleased with the progress the United States has made since he shared that story in his 2008 book, but the Chinese certainly did not heed his call to stand idly by while we put our house in order. By 2019 the United States was in second place among the world's nations in renewable energy installations, with a capacity of around 264.5 gigawatts. But that's an accomplishment that must be put into context. China, in first place, had a capacity of around 758.6 gigawatts.[118] They, along with many countries around the world, are leading the way, and the United States, for all of the progress made, has only been closing the gap.

Of course, countries are very different in population, energy needs, and resources, so to get an accurate scorecard of our progress in the clean energy race, it's more telling to think about the proportion of our energy that comes from renewable sources. Here, the story told by statistics is very different.

Iceland is the undisputed champion of renewable energy, a title they may never lose, because nearly 100 percent of their energy comes from renewable sources. Iceland's low population and unique resources make it possible to generate around 95 percent of the country's heating from hydropower and geothermal energy. And even with virtually no fossil fuel energy, Iceland generates nearly ten times the power per capita of the European Union.

Other top-ten countries on the list may or may not surprise you. Kenya produces 70 percent of its energy from renewables and aims to reach 100 percent in the next couple years. Scotland produces enough renewable energy to power all its homes and businesses without the need for any fossil fuels. And the United States, well, we may be a distant second in overall renewable power production, but we hold only eighth place worldwide by percentage of total energy use. Rolling into 2020, only 11 percent of energy produced in the United States, 17 percent of our electricity, came from renewable sources.[119]

So, though the momentum for renewable energy in the United States is strong, we have only begun to seize the strategic opportunities to lead the world in innovation, economic growth, and environmental stewardship. It pains me to admit that one of the main reasons we have struggled as a nation to fully capitalize on these opportunities as aggressively as other nations is because of our politics.

This is a point I want to make very clearly to Americans who identify as free market conservatives, as I do. The reason that the Republican Party has increasingly opposed a national commitment to clean renewable energy is not because conservatives are opposed to the pivot from fossil fuels on principle. As Devin Hartman put it in his 2018 *National Review* article, "A Conservative Energy Reset," "principled conservatives recognize that simply picking different winners than the left would pick is not conservative, it's just playing the big-government game in reverse. There's nothing conservative about coal and nuclear power, nor anything liberal about renewables and energy efficiency."[120]

He's right. There is widespread and growing support among Americans of every political persuasion that clean energy should become a national priority. The problem isn't politics or principle, per se; the problem is money. The wealthy fossil fuel industry has been buying its influence with conservatives for decades. Consider the following chart tracking the flow of fossil fuel contributions by political party.

Fossil Fuel Contributions by Party

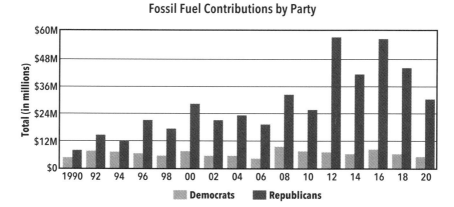

Source: Center for Responsive Politics. Oil & Gas, "Long-Term Contribution Trends [Table]," accessed on June 24, 2020, https://www.opensecrets.org/industries/totals.php?ind=E01++.

The record is clear. The fossil fuel industry has increasingly targeted one party, our party, as allies in its fight to delay the future of clean energy. These are not our principles, and in this case our party is not putting America's interests first. As Hartman puts it, "We do that by reducing regulation and encouraging competition, not by subsidizing the past."[121]

So as business leaders, entrepreneurs, citizens … Americans, it's time for us to set aside the foolish premise that a healthy, prosperous future based on clean renewable energy is a partisan issue. Clean energy stands as the greatest economic opportunity of this century. And it's not too late to be an *early* investor in renewable energy. You can, and so can the entire country.

A growing number of businesses already do. My medium-sized company saves $150 a day with our NZE HQ1. More than two million American homes and businesses now are solar powered, and the industry projects this number will double by 2023.[122] As more and more businesses realize that renewables can improve everything from brand value and balance sheets to productivity and profit and loss statements, we will see the birth and growth of whole new industries. To some extent, this is already happening. The question is whether your company will be a driver of the new energy economy—or be left in the dust.

Every business can play a role because every business uses energy, no matter what industry it is in. When you install solar panels on your roof, not only are you increasing the demand for renewable energy; you're increasing the supply at the same time. Renewable energy—solar, wind, geothermal—offers the opportunity to make every building a power plant. Imagine how much healthier our economy would be if companies produced wealth inside their factories and offices, on their roofs, and under their landscaping. It makes too much sense not to happen.

The undeniable economic logic is the reason I believe clean energy is the new gold rush of our time, even as some in our government struggle to see it. Energy is a sector ripe for change, where the best ideas—the ones that supply the most reliable power at the lowest price and that are safe and healthy—will win. We don't yet know what all of those ideas will be or where they'll originate, but therein lies the business opportunity. Renewable energy is still looking for its Steve Jobs or Bill Gates—someone who can mass market sustainability, so it becomes not only convenient but cool, and then ubiquitous.

Maybe you can be one of those visionaries, even at a local level. To invest in conservation, efficiency, and renewable energy is to position your business and the economy for long-term success. The size of our national challenge should not deter anyone from taking action right at home, where they are, and in their own business. When you take steps to maximize energy efficiency or invest in producing your own solar electricity, you'll find that the benefits go far wider and deeper than the drop in your overhead. You'll be investing in a change in your company's culture and in your community, and such "energy" tends to reflect back to you in unexpected ways. And your new "green and clean" brand doesn't need to be kept secret.

At Melink Corporation our commitment to clean energy is front and center to who and why we are. Customers like doing business with companies who sell more than a product or service—they want to buy from companies that are ethical, sustainable, and about making the world a better place. When families visit the Cincinnati Zoo, they park under a structure with a sign that heralds the Melink Solar Canopy. How many more customers could you reach if you sponsored a solar installation at a sports stadium? An amusement park? A mall where your products are sold? How far would that go, not only in promoting solar energy but your company's brand? Companies with the vision to lead the clean energy economy—those with the moxie to make it happen—will be the ones to benefit.

All others will wish they had.

PART 4
LEADERSHIP AND SUCCESS IN THE TWENTY-FIRST CENTURY

CHAPTER 12

Our Lives, Fortunes, and Sacred Honor

Posterity, you will never know how much it cost the present generation to preserve your freedom. I hope you will make good use of it. If you do not, I shall repent in heaven that ever I took half the pains to preserve it.
—JOHN ADAMS, IN A LETTER TO ABIGAIL ADAMS, APRIL 26, 1777

I f John Adams and his fellow giants who founded the American experiment were here to look back over the past two and a half centuries and judge how their posterity has fared, how would that assessment look, I wonder? I'd like to think that, in many cases, they would be satisfied—proud of how the American character has survived and even matured with every challenge. They would take note of all the times that America has risen to the defense of democracy around the world, helped to save and then rebuild modern Europe, reached out to nations afflicted by famine and authoritarian rule, and regularly defended itself against grave threats from within.

Of course, they would also perceive our many missteps, times when America fell short of her potential. They would shake their heads at the shortsightedness and greed of our manic business culture, where profits and wealth are held in higher regard than our ideals and values, like fairness, equality, and the common good. These ills were just as common in their time. In fact, they were so cognizant of the dangers of free markets without any anchor in morality that they crafted a constitution almost entirely occupied with protecting American citizens from the powerful and unscrupulous.

No, the quality that has set America apart through most of its history is not superior character; we've always had our fair share of both decency and villainy. In addition to our elegantly crafted constitution, we've depended on a bedrock of idealism more than any other country, the belief in the very idea of America. Without this indispensable element, we become less purposeful, less inspirational, and less exceptional. I'm quite sure that if Adams could speak to us today, he would celebrate the great nation we have been but also sound a warning about how we navigate the intersection at which we find ourselves. We must guide the invisible hand of the free market so that it follows more closely God's hand, he would say, hoping to pin our quest for an abundant life to higher ideals. Adams would tell us that wealth without wisdom and power without purpose are the only poisons that can threaten the American experiment. We cannot continue to lead and thrive as Americans if our progress as individuals and organizations is detached from our ethical and moral heritage. If so, we become something else, something less.

Let's not do that. Here we stand, at the threshold of the next great

American era. The United States remains the world's largest economy and possesses the most fearsome military might in history. This is no time to get complacent and take for granted the countless sacrifices that our forebears made for us. I think back to the words of President John F. Kennedy as he addressed the graduates of Amherst College in 1963: "I look forward to a great future for America—a future in which our country will match its military strength with our moral restraint, its wealth with our wisdom, its power with our purpose."[123] President Kennedy spoke these words at the height of the Cold War, when we both faced the peril of nuclear apocalypse and embraced the miraculous prospect of putting a man on the moon. Sixty years later, the president's words could not be more fitting—the contrasting threat and promise, starker. Life on this planet, including human life, faces the grave threat of climate change, and at the same time, the clean energy revolution offers us hope of a safer, healthier, and more prosperous future like the human race has never seen.

So moving forward, how do we stay true to the ideals of our founding fathers, face the great threat of our generation, and continue reaching for the stars? At this point, you won't be surprised by my answer. Incorporate *sustainability* in America's bedrock of core values along with freedom, equality, and justice. This is because sustainability represents all the good things we have just talked

> Sustainability is a natural extension of thinking and acting with integrity. It is being honest, respectful, and fair toward *everyone*—including future generations.

about. Wealth with wisdom; power with purpose. Sustainability is a natural extension of thinking and acting with integrity. It is being honest, respectful, and fair toward *everyone*—including future generations. And it is caring for our God-given natural world—the plant and animal life on which we all depend for sustenance and joy. Plundering our planet for short-term profit, on the other hand, is antithetical to integrity and sustainability. This is why I believe a heightened awareness and focus on sustainability by our leaders can renew faith in business and capitalism itself.

Our government can, and should, be an important driver of positive social change, but another aspect of the American character that I love is that we don't tend to wait for government to lead or provide all the answers. I think of Branch Rickey, general manager of the Brooklyn Dodgers, who decided the time was right to racially integrate Major League Baseball. When he put Jackie Robinson on his team in 1947, major civil rights legislation was still almost two decades away. Rickey signed Robinson because it was good business and the right thing to do. He could win more games with Robinson and other black players as well as establish a whole new fan base at the same time.

Rickey was right on both counts. The Dodgers that year started a run of six pennants in ten seasons. More important for the purposes of this discussion is that within a decade or so almost every team in the majors featured black players, and the game was better for it. The Dodgers, as the early adopters, gained the most.

Thanks to Rickey and Robinson, baseball's norms were changed. And a new normal in what was the country's most popular sport helped change America. Before we could pass laws banning discrimination, a critical mass of Americans had to decide prejudice was wrong. The private sector—in this case, professional sports—catalyzed change by leading in a very visible way.

TRANSCENDENT LEADERSHIP

Capitalism. It has both good and bad implications, depending on who you are talking with. But, regardless, going forward we should always strive to make it better, continuously improve it—like we do our products and services, our companies and business systems, and our government policies and institutions. To suggest we have perfected our economic system of capitalism is to ignore all the inequities, waste, and discord that can come from greed when it is left unchecked.

In the spirit of living up to our ideals, shouldn't we at least try to achieve something greater than saying this is the very best we can do?

Let me start with the premise that capitalism is—or should be—about competition, not greed. The difference might be subtle, but it is critical. This is because competition connotes and breeds positive, constructive behaviors, while greed connotes and breeds negative, destructive ones. No one questions when an engineer develops a better product for the market, or a salesperson builds a more trusted relationship with the customer. But everyone should question when an engineer cuts costs at the expense of safety, or a salesperson concocts an unsupportable pyramid scheme to earn higher commissions.

In other words, the unlimited potential of capitalism needs guardrails, just like the unlimited potential of freedom needs a constitution with amendments and laws. These guardrails are often decried as excess regulation, but without any regulation at all, we have societal anarchy and continual degradation in our norms. Sensible regulation protects against dangerous products and services entering the marketplace, prevents abuses of child labor and human trafficking, and ensures reasonable oversight of our health and environment.

It's said that sustainability is a three-legged stool encompassing the three Ps: profit, people, planet. No enterprise can be called sustainable unless it meets all three criteria. What's more, the sustainability stool falls unless all three legs are strong. Those who think and act on this level, in my opinion, deserve more than the distinction of being visionary leaders. I think of them as *transcendent leaders*. They are not only concerned about themselves and their companies. They are also committed to doing the right thing for their country and the world at large. This kind of leadership transcends the normally accepted "what's in it for me" mentality and promotes a more powerful "how can I help" ethic that gives true hope and meaning to a better future—one that is more fully aligned with America's founding values.

The most talented CEOs are able to imagine a future others don't yet see and have the perseverance and drive to make it happen. But too often their gaze is laser focused on increasing market share and profits to the exclusion of the other two legs of the stool. When their vision expands to include the social and environmental aspects of sustainability, something greater and more powerful happens. Their employees recognize and respect the higher values with which decisions are now made. And this promotes both productivity and pride in the greater cause toward which the company's mission is directed. Moreover, their customers recognize and respect these higher values, and this promotes stronger loyalty and longer-term engagement with the company. In short, the company's brand becomes more valuable, and the prospects for greater success in the future grow.

I shared my own experience at Melink Corporation, how a true commitment to all three legs of sustainability boosted our productivity and sales, expanded our market share and profitability, and burnished our brand for even greater future success. All around the country, around the world, businesses large and small are finding the same powerful benefits of broadening their vision beyond short-term profits.

Tom Murray of the Environmental Defense Fund, who advises companies including Walmart, McDonald's Corporation, and the

Procter & Gamble Company on reducing emissions, identifies a critical turning point: "We've moved past this concept that business versus the environment is a trade-off."[124] Instead, what companies around the world have discovered is that not only will their customers reward them with loyalty for operating sustainably, but more often than not doing the right thing also improves their bottom lines. United Airlines Holdings, Inc., for example, has been finding creative ways to make its planes lighter to improve fuel efficiency. The results are nothing less than astonishing. Not only are United aircraft pushing less carbon into the atmosphere; simply using a lighter paper stock for the in-flight magazine is also saving the company almost $300,000 per year on fuel. Add lighter beverage carts, redesigned bathrooms, and ending duty-free sales, and the airline has already saved over $2 billion in fuel costs.[125]

Apple, Google, Amazon … all of these tech giants have committed to powering their operations with renewable energy. Apple Park, Apple's new headquarters in Cupertino, California, is now the largest LEED Platinum–certified office building in North America. It is powered by 100 percent renewable energy from multiple sources, including a seventeen-megawatt on-site rooftop solar installation and four megawatts of biogas fuel cells, all controlled by a microgrid backed up by battery storage.

In other words, there does not have to be a conflict of sustainability and clean energy vis-à-vis profitability and growth. In fact, increasingly, if you don't adopt the three-legged stool approach to business and invest in all stakeholders, not just your investors, the market will punish your company for its selfish and short-term mindset. Sure, we are accountable to shareholders. But in the twenty-first century, we are also accountable to our employees, customers, vendors, and our community. And even, increasingly, to our nation and the world.

So back to the three-legged stool of sustainability. Profit is the easy one, since this is the metric businesspeople have been conditioned to maximize—often at the expense of the other two *P*s. However, disrespecting the other two *P*s, people and planet, are symptoms of the same disease of shortsightedness—and we can't treat one unless we are concerned about all three. Both stem from a cramped and shortsighted definition of self-interest. Both are a result of counting the profits that accrue this quarter while ignoring the bill that will eventually come due.

More to the point, a failure to think about sustainability underlies the difference between employment and exploitation. I don't mean employment in the usual economic sense but rather in a more general sense of making use of something. The difference in the dictionary definition of the words *employment* and *exploitation* might be only in their connotations, but in the real world, there is an enormous difference. Most people want to be employed; nobody wants to be exploited. A person who is exploited has no choice or only terrible choices. A person who is employed is part of a mutually beneficial arrangement and gains not only financially but also in social standing and self-esteem.

When it comes to energy, exploitation is what we've mostly done up to this point. When we burn fossil fuels, we are spending Earth's capital. By using energy built up over millions of years in just a few decades, we are bequeathing to future generations a planet that is hotter, poorer, and more dangerous than the one we were given. We get the benefits of cheap energy; they get only the hangover—that is precisely an exploitative relationship. Another way of saying that is we are spending our children's inheritance.

The bottom line is, when it comes to making our companies, our country, and the world more sustainable, we as leaders need to step up. The clean energy solutions are available and cost effective today ... we don't have to wait until 2025 or 2030 or whenever. Based on my experience, all we really need is the determination to implement them.

So I urge CEOs and every US leader to take the bold and necessary step of committing a percentage of their profits or capital budget—say, 5–10 percent—toward energy efficiency and renewable energy. Many companies are already doing this and not only save energy and money, becoming more sustainable and resilient, but also improve their culture and brand for long-term growth and success.

And the cumulative leadership of thousands of leaders and organizations doing this will usher in the clean energy revolution. This, in turn, will propel America to new heights in the decades ahead and improve not only our economy but also our security, health, and environment.

I believe the twenty-first century is a time of enormous challenge and opportunity for America. The question is, as a nation built upon the ideals of free enterprise and independence, do we choose a future of greed, short-term interests, and dependency ... or a future of integrity, long-term growth, and prosperity? Do we choose a future like so many world empires of the past, a future of decline and waning relevance, or will America

> **As a nation built upon the ideals of free enterprise and independence, do we choose a future of greed, short-term interests, and dependency ... or a future of integrity, long-term growth, and prosperity?**

turn the page and win the clean energy race of the twenty-first century? There is no reason we cannot choose renewal over decline, and to keep faith with our duty as American citizens and business leaders, we must.

It's time to return to our uniquely American roots, to the aspirational vision and values that John Adams and our founders hoped we would stay true to. But are we willing to pledge our lives, fortunes, and sacred honor like John Adams and our founders did, along with countless other American patriots and heroes over the approximately 250 years since our grand experiment was so daringly started? Or are we too comfortable, lazy, and perhaps even indifferent?

Love of country and civic duty occasionally call for sacrifice. They also call for doing the right thing. These are appropriate expectations for being lucky enough to call ourselves Americans. So we must never take for granted our freedoms, our health, and our environment in the face of current and future threats. And based on the trajectory and irreversible nature of climate change, this stands to be our greatest challenge—and opportunity—of all.

Let's make sure the legacy we leave is one that will make future generations proud of how we changed the course of history. For posterity.

CONCLUSION

A Conservative Platform for Uniting America

I t was our first family reunion in a couple years, and my siblings were just starting to wake up at the large bed and breakfast where we'd gathered. Most of them had arrived the evening before but several not until late in the night. I'd brewed a big pot of coffee and was still standing at the counter, munching on some toast and scanning the paper, when my sister Janine sneaked up from behind and threw her arm around me.

"Good morning, Steve!"

"Hey, Janine. How are you?" I asked with a smile, not surprised.

"Oh, so awesome. Isn't it great to have everyone together again?"

"It sure is, and I'm really glad that we've all stayed so close since Mom died … despite the distances that separate us."

"Yeah," she said, pausing for a moment before changing the subject. "So, Steve, how's life treating you?" Janine's always been the spiritual one in our family and loves to get to the heart of what matters.

I opened my mouth to answer her question, but before I could say a word, my brother Chris walked into the kitchen and blurted out, "Good morning, sibs! Coffee. Excellent!" My brother has a talent for dramatic entrances. "What, you didn't drive your fancy little golf cart down to the doughnut shop?" He stopped to hug Janine and reached his hand out to me as he went for the coffee pot with the other. "Well, I forgive you this time."

"You have a big heart, Chris," I chuckled.

"So, word is you're about to christen your second supergreen headquarters. How's that going?"

"Thanks for asking. Yeah, it's really going well. Our company's been fortunate to see continued growth since we built our first LEED building, and in addition to needing more space, we just wanted to show that NZE buildings are even more affordable today than our first HQ was fifteen years ago. I tell the team, 'If we're going to talk the talk … we need to walk the walk,' you know?"

Janine smiled and rested her hand on mine. "I think that's great, Steve. The world needs more innovation and leadership when it comes to clean energy in these crazy political times. We're proud of you."

Chris echoed the sentiment as Mary Frances stepped in and caught me on the cheek with a kiss. "Hello, you two," she smiled. "I'm pretty proud of this guy too."

It's hard to describe how much this affirmation meant to me, the support of my family.

"I'm just grateful," I replied. "It's an extension of my faith as much as anything. I hope you guys can visit soon and check it out!"

As my other brothers and sisters eventually drifted into the kitchen to get their coffee, the conversations naturally flowed to them, their families, and catching up on all of our busy lives. As I listened, my heart was filled with gratitude for them and for the encouragement they've given me over the years.

After breakfast I slipped out and went back to our room and the quiet of my mind. With the early sunlight beaming through the window illuminating the room and a small desk by the window, I sat, pulled a legal pad out of my bag, and began writing some thoughts down.

First, I reflected on how lucky I am. My wife is my best friend and soul mate. My children inspire me to be a better person. And my brothers and sisters are an extension of who I am in many ways. We shared the same parents, grew up in the same household, and laugh today about the many challenges of growing up in a family of nine. Family is core.

Second, despite our separate paths as adults living, working, and raising a family in different parts of the country, we all have a deep and abiding faith in God. Certainly, there are differences in how we practice our faith, but we all recognize that love transcends everything else in life. Faith gives us purpose.

Third, we all hold great respect for our country and the values on which it was founded. We all work in different professions and industries but believe in the power of American democracy, opportunity, and decency. And we all believe that, with hard work and ingenuity, we have a chance at the American Dream.

Fourth, the older we get, the more we appreciate Mother Nature for her inspiring beauty and positive effect on our health and happiness. We've all had access to good education and agree that science and technology should not be at odds with nature but work in concert with it. Our environment and the diversity of life that depends on it are divine gifts from God—and man should be their steward.

"These themes have been powerful influences in my life and the lives of my extended family," I wrote, "and they are wholly consistent with what most people would say are conservative, traditional values."

And by logical extension, the clean energy movement sweeping the world is wholly consistent with these themes too. After all, doing the right thing for our children, our country, and the world is fundamental to our survival as a species. It's not just about us, and it's not just about the present.

"Then … why?" I asked aloud in the room by myself. "Why has clean energy been reduced to a political issue in this country? And why have American conservatives taken the opposite position most people would expect given the core values they supposedly hold dear? It seems hypocritical … not to mention politically foolish given where voter preferences on this important topic are trending."

I continued writing, realizing this contradiction had been weighing on me for some time; I had to think this through.

Conservatism is about appreciating our freedom and the opportunity to pursue our dreams. It is about defending and fighting for truth, justice, and liberty. It is about being fiscally responsible for today, tomorrow, and the future. And it is about promoting a culture of life over death, hope and optimism over defeat and despair, and faith over indifference.

Clean energy is perfectly aligned with these values. The reasons why flowed freely from my pen:

- *It will create millions of jobs, thousands of companies, and a new clean economy.*

- *It will free us from our dependence on foreign oil and gas.*

- *It will reduce the power of our enemies, who sell this oil and gas to foment terrorism.*

- *It will make us healthier and more productive in a globally competitive world.*

- *It will make us more resilient with a distributed and smart power grid.*

- *It will improve our environment and reduce the costs and risks of climate change.*

- *It will save thousands and possibly millions of lives over the coming decades.*

I thought back to that LEED conference in Cleveland. The image of Kevin Hydes's last slide of his young daughter on the beach was seared in my memory. And his commitment to leaving her a better world reminded me to add one more to my list:

- *It will do right by our children and grandchildren.*

"All these strategic advantages, improvements to our quality of life, and ethical imperatives deserve the support of every American citizen and person of faith," I wrote. Imagine the power of Americans uniting as a country around this epic opportunity like we did to achieve victory in World War II. Imagine the transformative impact this would bring to people of every demographic. Like curing cancer and defeating terrorism, there can be no legitimate reasons *not* to rally around clean energy.

Then I reminded myself of what I had learned over the last ten years. The only reasons are bad ones: greed and politics. The fossil fuel industry brings essentially unlimited resources to bear on influencing uninformed and sometimes corrupt politicians in Washington and state capitols. And these politicians then obligingly follow the fossil fuel industry agenda and legislate in its favor.

I dropped my pen and leaned back in my chair, reading over what I'd just written. *And part of it is simply the Al Gore and Barack Obama factor*, I thought. *Politics is rarely about seeking the truth and almost always about opposing the other party. And since the most prominent proponents of making the transition to renewable energy an urgent national priority have been Democrats, it's natural, though ignorant and prideful, for most Republican politicians to instinctively stake out the opposite position. Even to this day, when coal plants are shutting down left and right, and even oil companies can see the writing on the wall, many conservative politicians are sticking with the party line that fossil fuels are our future. How backward, wrong, and sad.*

Perhaps the only way for stuck-in-the-mud politicians to see the facts for what they are is for citizens like me to disrupt this silly, partisan equation, I thought. *And, as a businessman, I can make a bigger impact by influencing people's pocketbooks than by influencing their politics.*

Intrigued with this thought, I went back to scribbling notes on that legal pad. "Our way of life is often more determined by our economy than any particular administration, which can change every four years. And, therefore, sustainable, fundamental change is ultimately driven by consumers—people like us."

While I was pretty sure that my family and friends would agree on the importance of consumer behavior to drive change, I had to admit that a relatively small percentage of those closest to me are actually doing this. We might all say the same things about the need for clean energy, but not everyone is putting their money where their mouth is.

In fact, polling data consistently shows that most Americans agree that climate change is a problem and that we need more clean energy. But what are most people doing about it?

Of course, I rationalized, they aren't in the clean energy business like I am. It's more important for business leaders like me to lead on this issue than it is for the individuals and families. And besides, the problem of climate change is too big for any one person to affect. That is why we need conservative political leaders to course correct.

I was going in circles, I realized, first assigning responsibility to government, then to business leaders like myself, and then to the average consumer. Pointing fingers was not going to accomplish anything. It then became clear to me the scale of change needed.

"*Everyone needs to be part of this change*," I said aloud and wrote across the top of my notes in bold letters. Otherwise, we are *all* part of the problem, not just the fossil fuel industry! Being passive about this is no longer acceptable. Just like being passive about racism, sexual harassment, age discrimination, and so many other social injustices is no longer acceptable.

> **Being passive about this is no longer acceptable.**

The fossil fuel industry and gas/electric utilities will sell us whatever we are willing to buy. And right now, most everyone in the United States is still buying coal-fired power, natural gas, and gasoline because they are what we're used to, and up until recently they've been the cheaper option. But just as we've begun to demand higher-quality foods that are fresh and devoid of chemicals, we need to create the demand for clean, sustainable, and domestic power from our energy suppliers.

"This is kind of like the drug problem in our country," I mused as I got up and walked around the room. "The fossil fuel industry and utilities might be the pushers, but we're the users. No, we're the addicts. They're creating the supply, but only because we're creating the demand. We need to stop creating the demand!"

I looked out the window and saw the evidence right there in the driveway in front of the house. It was filled with cars, SUVs, vans, and trucks that take in fossil fuels from gas pumps and hoses that deliver the drug like any other needles and syringes. And they belonged to my brothers and sisters and extended family.

My immediate family was really no different, I had to admit. Mary Frances drives a Lexus, my oldest son, Matt, drives a Toyota SUV, and Katie and Laura drive hybrids. Sure, my oldest daughter, Monica, has joined me in the "Tesla Club" but, to be honest, only because she just started working at my company, which provides an incentive for EVs.

The same careless consumerism exists with our friends and in our school and church communities. Most drive conventional cars and SUVs with internal combustion engines—without thinking for a second about the cumulative, adverse effects on our health and environment, not to mention our security and economy. You rarely see an electric car in our town, maybe a Chevy Bolt here and a Tesla there. Coworkers at our company are the exception, but only because we offer a nice incentive to buy EVs. What if we didn't? I am sure our company parking lot would look the same as every other parking lot in America.

I sat back down, wondering if I was being too critical—too judgmental. It's just their cars, how they get around. It's not like my family's choice between Tesla and GM is going to determine our global future. They're both great American brands, born of different generations.

But no, I thought. Electric cars have everything to do with clean energy. In fact, they're the perfect metaphor for the much larger clean energy movement. Cars are a symbol in America of who we are and what we stand for. They are an extension of our personal values and tastes, even more than our homes in many ways. In fact, electric cars are the closest things to a "consumer product" that we all want or need that allow us to directly participate in the clean energy movement.

But it just starts there. There are many other choices we can make besides cars to help usher in the clean energy revolution, and to the degree we can, we should. LEDs or florescent lights? Geothermal heating and cooling or air-source heat pumps? Solar or wind power? Home batteries or grid storage? Roof insulation or double-pane windows? The list of options for home and building owners goes on and on.

At that moment, Mary Frances came into the room. "What are you doing, Steve?" she asked. "Everyone is getting ready for a walk."

"Oh, I was just making a few notes."

And with my mind still deep in thought, I looked up and smiled at her.

"Mary Frances, what do you say we buy you an electric car next weekend? I know we've talked about this before, but this is a matter of principle on which I no longer want to send mixed signals to our kids and extended family and friends. Will you walk this walk with me—together?"

* * *

These conversations and thoughts have been played in my mind again and again like a Top 40 song since our family reunion. The alignment of clean energy with traditional values and the parallel contradiction in recent conservative political discourse. The world betting on technologies of the future and American conservatives betting on the extraction methods of the past. Utilities supplying brown power when most consumers want green power. And family and friends wanting to be part of the solution but not knowing how.

I have since concluded that we as conservatives need a complete overhaul of our perspective and positions on the increasingly important issue of energy. We need a platform that blows up all the false narratives of the fossil fuel industry and embraces clean energy as the way of the future. It truly is the way of the future, it's the fastest-growing sector of our national and global economy, and it's just getting started. Solar power and its other clean energy derivatives like wind and hydropower, as well as biofuels, are forever free.

I have no doubt that American capitalism is the best system to bring these clean energy technologies to scale. It has already proven to be the best system to feed, clothe, and shelter the billions of people we have on the planet. And while no system is perfect, capitalism unleashes our human potential—our passion, innovation, and productivity—like none other. Free markets with the proper guardrails are inherently more efficient and produce superior results compared to centralized, bureaucratic initiatives. That is, if the game isn't fixed by lobbying and disinformation campaigns.

It was through combining these two winning ideas—clean energy and capitalism—that I arrived at the concept of fusion capitalism. Solar fusion is literally the source of all the energy that sustains life on our planet. Capitalism is built on the premise that individuals are free to make their dreams come true in a free market of ideas, where the most useful innovations thrive. And when we consider the literal meaning of fusion (bringing or "fusing" together), we discover an even more powerful aspiration to pursue. After all, isn't our country and the world *starving* for—more than anything else—the opportunity to come together, and be united, in the quest for our common good?

It is against this backdrop that I developed the following clean energy platform for conservatives, consisting of five simple principles. I believe that United States and world markets would welcome an institutional commitment to these principles by the Republican Party, because long-term commitments at the party level give investors a longer runway to plan future investments and minimize risks—mitigating the policy shifts that come every four years with a new administration. The result will be better predictability, stronger growth, and a clear national commitment to win the clean energy revolution!

A CONSERVATIVE PLATFORM FOR AMERICA'S CLEAN ENERGY FUTURE

1. **Clean energy will secure the future of our economy, security, health, and environment.** Because of these multiple strategic benefits, we believe clean energy represents the biggest opportunity of the twenty-first century and should become America's number one national priority.

2. **Our goal is to achieve a zero-carbon economy by 2040.** This long-term vision can be achieved by any number of technologies and solutions, including energy efficiency, renewable energy, battery storage, natural gas carbon capture, smart grids, and electric transportation. Zero carbon is our moon shot times ten.

3. **The primary timelines and strategies are as follows:**

   a. **2025**: Eliminate the use of coal by replacing the power generated by all US coal plants with renewable energy. Convert coal jobs to safer and healthier clean energy jobs in solar, wind, and hydropower as well as biofuels. Also allow net metering and virtual net metering in all states.

   b. **2025**: Achieve a zero-growth carbon economy, meaning total carbon emissions must flatten and start to decline, even as we continue growing our economy. We must be a leader for the rest of the world on how to do this.

c. **2030**: Eliminate the use of oil and natural gas by electrifying the building and vehicle transportation sectors and replacing with renewable energy—mostly at the distributed level. Existing technologies can get us there; no silver bullets needed.

d. **2040**: Achieve 80 percent renewable energy across our economy. This will require distributed and grid-level battery storage for critical infrastructure. It will also require smart homes, buildings, and grids so that we have reliable power 24-7.

e. **2050**: Eliminate the use of "first-option" natural gas for all sectors, in addition to the building and transportation sectors, including utility power generation and industrial manufacturing. Natural gas in these other sectors can only be used for backup generators.

4. **Government should immediately eliminate subsidies to the fossil fuel industry and over five years eliminate subsidies to the clean energy industry.** Creating a level playing field will achieve most of the desired results. It will also put a price on carbon commensurate with its societal costs and make it tax-revenue neutral.

5. **Government may fine companies breaking these laws as necessary to ensure these goals, strategies, and timelines are met.** It may also require the fossil fuel industry to pay for public-awareness campaigns highlighting the dangers, costs, and risks of climate change.

Every new platform should, of course, invite scrutiny on how it will be paid for and who will pay for it. I want to reiterate that any incentive costs should be made tax-revenue neutral. In addition, I propose that the Office of Management and Budget (OMB) and General Accounting Office (GAO) develop the systems to track added tax revenues and savings over the years and decades based on the following:

- GDP growth, job creation, reduction of national debt, reduction of trade imbalance, and the like

- Reduced military deployment costs in the Middle East and in other parts of the world deemed unstable because of fossil fuel wars and terrorism

- Reduced military deployment costs for natural catastrophes, rising sea levels, and mass human migration

- Increased health and productivity of America's students and workers; less respiratory and heart disease, lower medical costs and absenteeism

- Reduced federal emergency costs associated with wildfires, storms, hurricanes, and sea level rise

- Reduced insurance costs

I believe these revenue gains and savings will amount to trillions of dollars in the coming years and decades, far above the platform costs themselves, which, again, should be paid for by revenue-neutral means.

Conservatives would do well to embrace this platform for three compelling reasons: (1) We'll be obliged to anyway as economic forces more powerful than politics continue moving in this direction, (2) the Democratic Party will beat us to the punch with their Green New Deal or some variant of it, and (3) it's the right thing to do.

I say we move now for the third reason ... before the first and second reasons make us look like the Out-of-Touch-with-Reality Party we otherwise risk becoming. Even the conservative Vatican is now urging Catholics, and all people, to stop investing in (and using) fossil fuels. This has become one of the moral issues of our time.

Conservatives would be much better served in every way by engaging the Democratic Party on how to address the issues of climate change and renewable energy rather than denying the former and stymying the latter. We can work together on this number one national priority without agreeing on every detail of the best way to move forward. Rather than pitting ourselves against each other, wasting time, energy, and resources, let's win this war together as a united people, as the United States. Sure, there will be details to be worked out by members of the House and Senate, but let's show the world the power of real leadership when our future is on the line. Our leadership over the next ten to twenty years can and should be our greatest legacy.

Most importantly, this proposed platform is simple, bold, cost effective, and true to conservative values! And it would be a unifying stroke of genius for every Republican politician advocating for it. Imagine the goodwill and swell of support that would come from such a battle cry of high purpose. Even a strategist focused solely on winning votes has to acknowledge this is a vote-winning position for the right. Americans and people the world over would celebrate and reward a shift away from the past toward a clean, healthy, moral future.

Moderate voters, who often feel abandoned by the extreme positions taken by the left and right, would have a new reason to consider the conservative agenda. After all, clean energy and its many benefits are extremely popular with most Americans. In 2017 a PEW poll found that 75 percent of Americans want the government to give priority to solar over fossil fuels. This is an opportunity for the

Republican Party to gain a new lease on life! This isn't some kind of concession to the Democrats or the left. This is about taking the best of what we know from economics, science, and history and building a coalition of leaders across the political spectrum to craft forward-thinking legislation and policies to save the world and make it a better place at the same time. Thoughtful, decent people all like that.

The real questions are: Can leaders within the Republican Party rise to the occasion and do the right thing despite the party's recent positions on these issues? Can we reach out to moderate and even progressive voices in the spirit of compromise and magnanimity—true signs of government leadership? Can we embrace our party's longer and principled history on the environment with pride and look to it for inspiration? And can we help educate and inspire our base to commit to a revival going forward?

All Americans ... and the world ... coming together. Finally.

This is the promise of fusion capitalism for the twenty-first century. It's not about division and the politics of things—guns, walls, and flags. It's about respect and the politics of people—ourselves, our children, and future generations.

> It's not about division and the politics of things— guns, walls, and flags. It's about respect and the politics of people— ourselves, our children, and future generations.

It's not about the old guard getting its way. It's about the majority of people of all colors and backgrounds being the wonderful melting pot we call the United States of America. It's not about going right or left. It's about going *forward* with a shared sense of national purpose and civic duty.

Let me leave you with an acronym for FUSION that helps codify fusion capitalism in easy-to-remember terms. Terms that conservatives must embrace to win the hearts and minds of an increasingly diverse voter population:

F = Future. It also stands for *faith* and *family* ... as well as *first*.

U = United. Our country was named the *United States* for a reason.

S = Solar. It also stands for *science* and its offspring, new technologies.

I = Innovation. It also stands for *international* and *inclusion*.

O = Opportunity. We must lead the greatest *opportunity* of the century, or others will.

N = Nature. We should respect Mother Nature and life as God's ultimate gifts.

Who am I to share my story and propose such a big vision and platform?

I am a man of faith, a US and world citizen, a husband and father, an engineer and businessman. My politics do not singularly define me, but I generally subscribe to conservative values. Having said this, I believe the Republican Party has been wrong on several issues, most notably climate change. It's time to get on the right track.

As someone who's built his career and company on the laws of physics and engineering sciences, I do not approach the issue of climate change from a political perspective. I approach it from a safety, cost, and risk perspective. In fact, in my real job, if I don't, my work can lead to failures and even catastrophes. As well as lost jobs and lives.

Additionally, I approach the issue of climate change from the perspective that if our company can build two NZE headquarters for our growing business, then other individuals and companies can too. The technologies are proven and financially sound. And if we can make a business case for supercool and reliable electric cars, then most anyone can. This isn't pie-in-the-sky stuff.

Ultimately, supporting life and what I believe is God's purpose for all of us is what drives me. And since climate change ranks with nuclear war and disease as one of the greatest threats to our human race, I'm prone to action in the area I know best. Let's not throw away our history, future, and possibly our souls because of indifference. As Pope Francis recently said, "The only greater tragedy that could come from this pandemic of disease is if we *return* to a pandemic of indifference."

My life journey is just a microcosm of our country's journey. By finding purpose, my company and I became stronger and more successful. I believe America must find its purpose again too. We are wealthy, we are educated, and we are the most innovative society in the world, but we need to apply these gifts to a higher purpose than just making more money. Our country and the world desperately need our leadership, because the future of mankind and our planet hangs in the balance.

The good news is, the clean energy revolution is the prosperous path, it is the moral path, and it is a particularly American path.

Fusion capitalism … is our destiny.

POSTSCRIPT

t doesn't hurt that electric cars, solar panels, and wind turbines are seen by the majority of people as cool and sexy. When I recently told Mary Frances that she'd look *hot* in an electric car, she paused a moment, smiled, and then asked, "Do you think I'd look okay in a Model S?" Our whole family is signing up for the next American revolution … one loved one at a time!

I hope you will join us and help lead the way.

ACKNOWLEDGMENTS

I have several people to thank for making this book possible.

First, I must thank my wife, Mary Frances. She has been a loving and steadfast supporter of my tightrope career and never wavered despite the many near-falls. My cup is always full with her by my side, and this has sustained and empowered me to do crazy things like start businesses and try to change the world.

Second, I owe my writing coach, David Allen, a debt of gratitude for his help with the storytelling process. He helped me communicate in such a way that readers will come to know me before taking any personal or leadership leap of faith with me. He is also a first-rate researcher, professor, and humanitarian.

Third, I must thank my publisher, ForbesBooks, for giving me a national/global platform and audience with which to share my story. Their cover and website design services, podcast program, speaker series, and blogging support are second to none, and their media relations expertise will serve me for a long time to come.

Last, but not least, I want to thank my employee-owners for inspiring me along this thirty-five-year journey. Any success I have enjoyed is largely because of them. Their dedication to our vision, mission, and values has made working at our company a privilege and honor. I look forward to the exciting sequel they will write in the years ahead.

ABOUT THE AUTHOR

S teve is the founder and CEO of Melink Corporation, a pioneer and leader in energy efficiency and renewable energy solutions for the commercial building industry since 1987. The company became employee owned in 2018.

Based in Cincinnati, Ohio, the company comprises five growing businesses: HVAC commissioning, kitchen ventilation controls, building health monitoring, solar PV systems, and supergeothermal HVAC systems.

Its customers include many of the largest and most successful companies in the world. Melink serves restaurant, retail, supermarket, and hotel chains as well as colleges and universities, hospitals and nursing homes, casinos and resorts, and military and government, among others.

Melink's corporate headquarters consists of two net-zero-energy buildings, both widely regarded as two of the greenest buildings in the United States. And its vehicle fleet consists of all hybrid and electric cars, with numerous charging stations at both buildings.

Steve is a national expert and speaker on sustainability, clean energy, and zero-energy buildings. He has consulted with federal and state legislators on multiple occasions and has testified before the state of Ohio legislature numerous times as well. He authored the book *CEO Power & Light: Transcendent Leadership for a Sustainable World* in 2015.

Steve and his wife, Mary Frances, have four children. He earned a BS degree in mechanical engineering from Vanderbilt University and an MBA from Duke University. Steve is an avid reader of US history, leadership, and all things sustainability.

ENDNOTES

1   Ross Geredien, "Post-Mountaintop Removal Reclamation of Mountain Summits for Economic Development in Appalachia," Natural Resources Defense Council, December 7, 2009, accessed May 28, 2015, http://www.ilovemountains.org/reclamation-fail/mining-reclamation-2010/MTR_Economic_Reclamation_Report_for_NRDC_V7.pdf.

2   US Environmental Protection Agency, "Outdoor Air——Industry, Business, and Home: Oil and Natural Gas Production—Additional Information," accessed December 6, 2014, http://www.epa.gov/oaqps001/community/details/oil-gas_addl_info.html#activity2.

3   Richard Perez-Pena, "US Maps Pinpoint Earthquakes Linked to Quest for Oil and Gas," *New York Times*, April 13, 2015, accessed April 14, 2015, http://www.nytimes.com/2015/04/24/us/us-maps-areas-of-increased-earthquakes-from-human-activity.html?ref=science.

4   Oklahoma Geological Survey, "Earthquakes in Oklahoma," accessed April 14, 2015, http://earthquakes.ok.gov/what-we-are-doing/oklahoma-geological-survey/.

5   Anthony J. Marchese and Dan Zimmerle, "The US natural gas industry is leaking way more methane than previously thought," The Conversation, July 2, 2018, accessed May 26, 2020, https://theconversation.com/the-us-natural-gas-industry-is-leaking-way-more-methane-than-previously-thought-heres-why-that-matters-98918.

6   "Environmental Impacts of Coal Power: Air Pollution," Union of Concerned Scientists, accessed April 28, 2015, http://www.ucsusa.org/clean_energy/coalvswind/c02c.html#.VT_kqJNUWgw.

7   Victor Lavy, Avraham Ebenstein, and Sefi Roth, "The Impact of Short Term Exposure to Ambient Air Pollution on Cognitive Performance and Human Capital Formation," National Bureau of Economic Research, October 2014, accessed April 28, 2015, http://www.nber.org/papers/w20648.

8   American Lung Association, "Asthma & Children Fact Sheet," accessed July 16, 2015, http://www.lung.org/lung-disease/asthma/resources/facts-and-figures/asthma-children-fact-sheet.html.

9   American Lung Association.

10   Taft Wireback, "The Day the River Turned Gray," (Greensboro) *News & Record*, February 1, 2015, 2.

11   (Greensboro) *News & Record*, "Coal Ash by the Numbers," February 1, 2015, 16.

12   Paul Hawken, *The Ecology of Commerce: A Declaration of Sustainability* (New York: HarperCollins, 1993), 21.

13   Intergovernmental Panel on Climate Change (IPCC), *Climate Change 2014: Synthesis Report*, Contribution of Working Groups I, II and III to the Fifth Assessment Report of the Intergovernmental Panel on Climate Change [Core Writing Team, R. K. Pachauri and L. A. Meyer (eds.)], 4, accessed November 28, 2014, https://www.ipcc.ch/site/assets/uploads/2018/05/SYR_AR5_FINAL_full_wcover.pdf.

14   Intergovernmental Panel on Climate Change, *Climate Change 2014: Synthesis Report*, 29.

15   IPCC, 40.

16   Climate Central, "Top 10 Warmest Years on Record," accessed May 26, 2020, https://www.climatecentral.org/gallery/graphics/top-10-warmest-years-on-record.

17   Rebecca Lindsey and LuAnn Dahlman, "Climate Change: Global Temperature, NOAA," Climate.gov, January 16, 2020, accessed May 26, 2020, https://www.climate.gov/news-features/understanding-climate/climate-change-global-temperature.

18   "Oil Industry," History, August 21, 2018, accessed May 30, 2020, https://www.history.com/topics/industrial-revolution/oil-industry.

19 US Energy Information Administration, "The United States is now the largest global crude oil producer," September 12, 2018, http://www.eia.gov/todayinenergy/detail.php?id=37053.

20 Knoema World Data Atlas, "Total Petroleum Consumption," accessed April 9, 2020, https://knoema.com/atlas/topics/Energy/Oil/Petroleum-consumption.

21 US Energy Information Administration, "What countries are the top producers and consumers of oil?," accessed April 9, 2020, https://www.eia.gov/tools/faqs/faq.php?id=709&t=6.

22 US Energy Information Administration, "How much oil consumed by the United States comes from foreign countries?," accessed April 9, 2020, https://www.eia.gov/tools/faqs/faq.php?id=32&t=6.

23 US Energy Information Administration, "The United States produces a large share of the petroleum it consumes, but it still relies on imports to help meet demand," accessed April 9, 2020, https://www.eia.gov/energyexplained/oil-and-petroleum-products/imports-and-exports.php.

24 Alan S. Blinder and Jeremy B. Rudd, "The Supply-Shock Explanation of the Great Stagflation Revisited," National Bureau of Economic Research, December 2008, 15, accessed November 29, 2014, http://www.nber.org/papers/w14563.

25 Binder and Rudd, 15.

26 Brian O'Keefe, "Chevron's Cheap-Oil Playbook," *Fortune*, April 1, 2015, 22.

27 US Energy Information Administration, "International Energy Outlook 2015," April 2015, accessed April 19, 2015, http://www.eia.gov/forecasts/aeo/pdf/0383%282015%29.pdf.

28 James D. Hamilton, "Historic Oil Shocks," National Bureau of Economic Research, February 2011, accessed April 19, 2015, http://www.nber.org/papers/w16790.

29   David Coady, Ian Parry, Nghia-Piotr Le, and Baoping Shang, *IMF Working Paper—Global Fossil Fuel Subsidies Remain Large: An Update Based on Country-Level Estimates*, May 2019, accessed June 1, 2020, https://www.imf.org/en/Publications/WP/Issues/2019/05/02/Global-Fossil-Fuel-Subsidies-Remain-Large-An-Update-Based-on-Country-Level-Estimates-46509.

30   Coady et al.

31   James Stavridis, "On Watch in the Arabian Gulf: What the US Navy Faces Against Iran," *Time*, June 25, 2019, accessed June 3, 2020, https://time.com/5613558/tensions-iran-u-s-sailors/.

32   "US Secretary of State, Terrorist Finance: Action Request for Senior Level Engagement on Terrorism Finance," *The Guardian*, December 30, 2009, accessed December 1, 2014, http://www.theguardian.com/world/us-embassy-cables-documents/242073.

33   Steve A. Yetiv, *The Petroleum Triangle: Oil, Globalization, and Terror* (Ithaca, NY: Cornell University Press, 2011), 6.

34   Marc Champion, "Who's Afraid of a Russian Gas Cut?," BloombergView, October 20, 2014, accessed December 1, 2014, http://www.bloom-bergview.com/articles/2014-10-20/who-s-afraid-of-a-russian-gas-cut.

35   Energy Policy Information Center, "The Growing Connection Between Oil and Terror," September 4, 2014, accessed December 1, 2014, http://energypolicyinfo.com/2014/09/the-growing-connection-between-oil-and-terror/.

36   Jim Michaels, "US coalition slashes ISIS oil revenue by more than 90%," *USA Today*, October 2, 2017, accessed June 3, 2020, https://www.usatoday.com/story/news/world/2017/10/02/u-s-coalition-slashes-isis-oil-revenue-more-than-90/717303001/.

37   Kate Fazzini, "The energy industry practices for a 'black swan' cyber-attack that could take down the grid," CNBC, November 16, 2019, accessed June 3, 2020, https://www.cnbc.com/2019/11/16/energy-sector-practices-for-a-black-swan-cyberattack.html.

38 Rachel Frazin, "What to know about cyberattacks targeting energy pipelines," The Hill, March 1, 2020, accessed June 3, 2020, https://thehill.com/policy/energy-environment/485254-what-to-know-about-recent-cyberattacks-on-energy-pipelines.

39 Frazin.

40 Christine Buurma and Alyza Sebenius, "Cyberattack targets oil infrastructure, shuttering facility for two days," World Oil, February 20, 2020, accessed June 3, 2020, https://www.worldoil.com/news/2020/2/20/cyberattack-targets-oil-infrastructure-shuttering-facility-for-two-days.

41 "Summary of *Terrorism and the Electric Power Delivery System*," National Academies Press, accessed December 2, 2014, http://www.nap.edu/openbook.php?record_id=12050&page=1#ths1_1.

42 Niall McCarthy, "The Annual Cost Of The War In Afghanistan Since 2001," *Forbes*, September 12, 2019, accessed July 9, 2020, https://www.forbes.com/sites/niallmccarthy/2019/09/12/the-annual-cost-of-the-war-in-afghanistan-since-2001-infographic/#7c34a4a41971.

43 "Summary of *Terrorism*."

44 US Department of Defense, "2014 Climate Change Adaptation Roadmap," accessed April 29, 2015, http://thehill.com/policy/energy-environment/220577-read-dod-report-2014-climate-change-adaptation-roadmap.

45 Jonathan Gregory, "Projections of Sea Level Rise, Climate Change 2013: The Physical Science Basis," IPCC, accessed April 29, 2015, https://www.ipcc.ch/pdf/unfccc/cop19/3_gregory13sbsta.pdf.

46 Nancy Lord, *Early Warming: Crisis and Response in the Climate-Changed North* (Berkeley, CA: Counterpoint, 2011), 171.

47 Colin P. Kelley, Shahrzad Mohtadi, Mark A. Cane, Richard Seager, and Yochanan Kushnir, "Climate Change in the Fertile Crescent and Implications of the Recent Syrian Drought," Proceedings of the National Academy of Sciences of the United States of America," January 30, 2015, accessed April 23, 2015, http://www.pnas.org/content/112/11/3241.

48 "Ensuring America's Freedom of Movement: A National Security Imperative to Reduce US Oil Dependence," accessed December 1, 2014, CNA Corporation, http://www.cna.org/sites/default/files/MAB4.pdf.

49 First Congress of the United States, *Joint Resolution of Congress proposing the Bill of Rights,* The National Archives of the United States of America, September 25, 1789, accessed June 4, 2020, https://www.archives.gov/founding-docs/bill-of-rights-transcript.

50 Center for Responsive Politics, "Our Vision and Mission: Inform, Empower & Advocate," OpenSecrets, accessed June 4, 2020, https://www.opensecrets.org/about/.

51 Karin Kirk, "Fossil fuel political giving outdistances renewables 13 to one," Yale Climate Connections, January 6, 2020, accessed June 4, 2020, https://www.yaleclimateconnections.org/2020/01/fossil-fuel-political-giving-outdistances-renewables-13-to-one/#:~:text=During%20the%202017%2D2018%20midterm,than%20%24359%20million%20in%20two.

52 Kirk.

53 Union of Concerned Scientists, "How Fossil Fuel Lobbyists Used 'Astroturf' Front Groups to Confuse the Public," October 11, 2017, accessed June 6, 2020, https://www.ucsusa.org/resources/how-fossil-fuel-lobbyists-used-astroturf-front-groups-confuse-public.

54 Union of Concerned Scientists.

55 Union of Concerned Scientists.

56 Lynn J. Good, "A Message from our CEO," in *2019 Duke Energy Sustainability Report*, April 28, 2020, accessed June 8, 2020, https://sustainabilityreport.duke-energy.com/introduction/a-message-from-our-ceo/.

57 Environmental Working Group, *Public Energy Enemy No. 1: Why Duke, America's Biggest Electric Utility, Is Also the Worst for the Environment*, May 1, 2019, accessed June 8, 2020, https://www.ewg.org/energy/dukeenergy/press-release.php.

58   Environmental Working Group, *Public Energy Enemy*.

59   Scott Smith, "2020 Power & Utilities Industry Outlook," *Wall Street Journal CFO Journal*, February 13, 2020, accessed June 8, 2020, https://deloitte.wsj.com/cfo/2020/02/13/2020-power-utilities-industry-outlook/.

60   Ryan Sabalow, Ryan Lillis, Dale Kasler, Alexandra Yoon-Hendricks, and Phillip Reese, "'This fire was outrunning us': Surviving the Camp Fire took bravery, stamina and luck," *Sacramento Bee*, November 25, 2018, updated May 8, 2019, accessed May 14, 2020, https://www.sacbee.com/news/california/fires/article221980270.html.

61   Jonathan Watts, "Wildfires rage in Arctic Circle as Sweden calls for help," *The Guardian*, July 18, 2018, accessed May 14, 2020, https://www.theguardian.com/world/2018/jul/18/sweden-calls-for-help-as-arctic-circle-hit-by-wildfires.

62   Carolyn Gramling, "Australia's wildfires have now been linked to climate change," *Science News*, March 4, 2020, accessed May 14, 2020, https://www.sciencenews.org/article/australia-wildfires-climate-change.

63   University of Waterloo, SciTech Daily, November 17, 2019, accessed May 14, 2020, https://scitechdaily.com/arctic-permafrost-turning-into-a-carbon-source-holds-more-carbon-than-has-ever-been-released-by-humans/.

64   Marshall Shepard, "You Don't Live In The Arctic But Climate Change There Affects You Too—Here Are 3 Reasons," *Forbes*, December 11, 2019, https://www.forbes.com/sites/marshallshepherd/2019/12/11/you-dont-live-in-the-arctic-but-climate-change-there-affects-you-toohere-are-3-reasons/#6f4d1599f0c8.

65   Elizabeth Kolbert, *The Sixth Extinction: An Unnatural History* (New York: Henry Holt and Company, 2014), 17.

66   Carl Zimmer, "Study Finds Global Warming as Threat to 1 in 6 Species," *New York Times*, April 30, 2015, http://www.nytimes.com/2015/05/05/science/new-estimates-for-extinctions-global-warming-could-cause.html?ref=science.

67 International Energy Agency, "World Energy Investment Outlook," 2014, accessed May 28, 2015, http://www.iea.org/publications/freepublications/publication/weio2014.pdf.

68 Jeffrey Kluger, "Senator Throws Snowball! Climate Change Disproven!," *Time*, February 27, 2015, https://time.com/3725994/inhofe-snowball-climate/.

69 Climate Central, *Global Weirdness: Severe Storms, Deadly Heat Waves, Relentless Drought, Rising Seas, and the Weather of the Future* (New York: Pantheon Books, 2012), 88.

70 "Scientific Consensus: The Earth's Climate is Warming, NASA, Global Climate Change Vital Signs of the Planet," NASA, accessed May 20, 2020, https://climate.nasa.gov/scientific-consensus/.

71 Dana Nuccitelli, "Trump thinks scientists are split on climate change. So do most Americans," *The Guardian*, October 22, 2018, accessed May 20, 2020, https://www.theguardian.com/environment/climate-consensus-97-per-cent/2018/oct/22/trump-thinks-scientists-are-split-on-climate-change-so-do-most-americans.

72 Olga Khazan, "Agent Orange's Health Effects Continued Long After the Vietnam War's End," *Washington Post*, August 9, 2012, accessed December 5, 2014, http://www.washingtonpost.com/blogs/worldviews/post/agent-oranges-health-effects-continued-long-after-the-vietnam-wars-end/2012/08.09/484d03cc-e243-11e1-ae7f-d2a13e-249eb2_blog.html.

73 Holly Morris, "After Chernobyl, They Refused to Leave," CNN, November 7, 2013, accessed December 5, 2014,http://www.cnn.com/2013/11/07/opinion/morris-ted-chernobyl/.

74 Sarah Zielinski, "Ocean Dead Zones are Getting Worse Globally Due to Climate Change," *Smithsonian*, November 10, 2014, accessed April 28, 2015, http://www.smithsonianmag.com/science-nature/ocean-dead-zones-are-getting-worse-globally-due-climate-change-180953282/?no-ist.

75   Kolbert, *The Sixth Extinction*, 6.

76   Climate Central, *Global Weirdness*, 25.

77   IPCC, *Climate Change 2014*, 48.

78   Justin Gillis, "New Study Links Weather Extremes to Global Warming," *New York Times*, April 27, 2015, accessed April 30, 2015, http://www.nytimes.com/2015/04/28/science/new-study-links-weather-extremes-to-global-warming.html?ref=science.

79   Justin Doom, "Antarctica Losing Mt. Everest's Worth of Ice as Melt Triples," Bloomberg New Energy Finance, December 2, 2014, accessed December 6, 2014, http://about.bnef.com/bnef-news/antarctica-losing-mt-everest-s-worth-of-ice-as-melt-triples-1/.

80   Lord, *Early Warming*, 142.

81   IPCC, *Climate Change 2014*, 11.

82   Jeffrey Goldberg, "Drowning Kiribati," Bloomberg Business, November 21, 2013, accessed December 1, 2014, http://www.bloomberg.com/bw/articles/2013-11-21/kiribati-climate-change-destroys-pacific-island-nation.

83   National Snow & Ice Data Center, "Methane and Frozen Ground," accessed May 28, 2015, https://nsidc.org/cryosphere/frozenground/methane.html.

84   Climate Central, *Global Weirdness*, 62.

85   Lord, *Early Warming*, 198.

86   Maggie Molina, Patrick Kiker, and Seth Nowak, "The Greatest Energy Story You Haven't Heard: How Investing in Energy Efficiency Changed the US Power Sector and Gave Us a Tool to Tackle Climate Change," American Council for an Energy-Efficient Economy, August 2016, accessed June 11, 2020, http://www.ourenergypolicy.org/wp-content/uploads/2016/08/The-Greatest-Energy-Story.pdf.

87   Molina et al.

88   William Nordhaus, *Climate Casino: Risk, Uncertainty, and Economics for a Warming World* (New Haven, CT: Yale University Press, 2013), 263.

89   Sameer Kwatra and Chiara Essig, "The Promise and Potential of Comprehensive Commercial Building Retrofit Programs," American Council for an Energy-Efficient Economy, May 2014, accessed December 16, 2014, http://aceee.org/sites/default/files/publications/researchreports/a1402.pdf.

90   Casey J. Bell, "Energy Efficiency Job Creation: Real World Experiences," American Council for an Energy-Efficient Economy," October 2012, accessed December 16, 2014, http://aceee.org/sites/default/files/pdf/white-paper/energy-efficiency-job-creation.pdf.

91   Kwatra and Essig, "The Promise and Potential."

92   Kwatra and Essig.

93   International Energy Agency, "Energy Efficiency Market Report 2014," Executive Summary, accessed December 14, 2014, https://www.iea.org/Textbase/npsum/EEMR2014SUM.pdf.

94   International Energy Agency.

95   Lovins, *Reinventing Fire*, 239.

96   Tom Randall, "While You Were Getting Worked Up Over Oil Prices, This Just Happened to Solar," Bloomberg Business, October 29, 2014, accessed November 24, 2014, http://www.bloomberg.com/news/2014-10-29/while-you-were-getting-worked-up-over-oil-prices-this-just-happened-to-solar.html.

97   Kathryn Parkman, "Solar Energy vs. Fossil Fuels," Consumer Affairs, March 31, 2020, accessed June 17, 2020, https://www.consumeraffairs.com/solar-energy/solar-vs-fossil-fuels.html.

98 Drew Desilver, "Renewable energy is growing fast in the US, but fossil fuels still dominate," Pew Research Center, January 15, 2020, accessed June 17, 2020, https://www.pewresearch.org/fact-tank/2020/01/15/renewable-energy-is-growing-fast-in-the-u-s-but-fossil-fuels-still-dominate/.

99 Vera Eckert, "Renewable energy's share of German power mix rose to 46% last year: research group," Reuters, January 3, 2020, https://www.reuters.com/article/us-germany-power-outputmix/renewable-energys-share-of-german-power-mix-rose-to-46-last-year-research-group-idUSKBN1Z21K1.

100 Destatis Statistisches Bundesamt, "August 2019: number of persons in employment up 0.7% year on year," August 29, 2019, accessed June 18, 2020, https://www.destatis.de/EN/Press/2019/09/PE19_384_132.html#:~:text=The%20adjusted%20unemployment%20rate%20was%203.1%25%20in%20August%202019.&text=Results%20of%20employment%20accounts%20as,in%20Germany%20(resident%20concept).

101 Bureau of Labor Statistics, US Department of Labor, "Unemployment rate unchanged at 3.6 percent in May 2019," accessed June 18, 2020, https://www.bls.gov/opub/ted/2019/unemployment-rate-unchanged-at-3-point-6-percent-in-may-2019.htm.

102 Zoe Casey, "Denmark: 50% Wind Powered Electricity by 2020," European Wind Energy Association (blog), July 16, 2012, accessed December 18, 2014, http://www.ewea.org/blog/2012/07/denmark-50-wind-powered-electricity-by-2020/.

103 "Denmark's renewable energy use passes landmark and is poised to grow in 2020s," The Local, January 23, 2020, https://www.thelocal.dk/20200103/denmarks-renewable-energy-use-passes-landmark-level-and-is-poised-to-grow-in-2020s.

104 Government of Denmark, "Independent from Fossil Fuels by 2050," accessed December 18, 2014, http://denmark.dk/en/green-living/strategies-and-policies/independent-from-fossil-fuels-by-2050/.

105 Danish Wind Industry Association, "Denmark—Wind Energy Hub," accessed December 18, 2014, http://ipaper.ipapercms.dk/Windpower/Englishpublications/Denmark_Wind_Energy_Hub/.

106 Danish Export Association, "Danish Wind Energy Group," accessed December 18, 2014, http://www.dk-export.com/networks/danish-wind-energy-group/.

107 T. Wang, "Distribution of solar photovoltaic module production worldwide in 2018, by country," Statista, February 10, 2020, accessed June 18, 2020, https://www.statista.com/statistics/668749/regional-distribution-of-solar-pv-module-manufacturing/.

108 Erin R. Pierce, "Top 9 Things You Didn't Know About America's Power Grid," US Department of Energy, November 20, 2014, accessed December 18, 2014, http://energy.gov/articles/top-9-things-you-didnt-know-about-americas-power-grid.

109 Sören Amelang and Freja Eriksen, "Germany's electricity grid stable amid energy transition," Clean Energy Wire, January 31, 2020, accessed June 18, 2020, https://www.cleanenergywire.org/factsheets/germanys-electricity-grid-stable-amid-energy-transition.

110 David Darling and Sara Hoff, "Average US electricity customer interruptions totaled nearly 8 hours in 2017," US Energy Information Administration, November 30, 2018, accessed June 18, 2020, https://www.eia.gov/todayinenergy/detail.php?id=37652.

111 "Investment tax credit for solar power," EnergySage, updated March 11, 2020, accessed June 20, 2020, https://www.energysage.com/solar/cost-benefit/solar-investment-tax-credit/#:~:text=The%20history%20of%20the%20solar,at%20the%20end%20of%20 2007.&text=2021%3A%20Owners%20of%20new%20 residential,the%20system%20from%20their%20taxes.

112 Beth Daley, "A tax increase that's proven to save lives," The Conversation, November 29, 2017, accessed June 20, 2020, https://theconversation.com/a-tax-increase-thats-proven-to-save-lives-87908.

113 Miranda Green and Alex Gangitano, "Oil companies join blitz for carbon tax," The Hill, May 22, 2019, accessed June 20, 2020, https://thehill.com/policy/energy-environment/445100-oil-companies-join-blitz-for-carbon-tax.

114 Henry Sanderson, "Clean energy shares streak ahead of fossil fuel stocks," *Financial Times*, October 1, 2019, accessed June 22, 2020, https://www.ft.com/content/2586fa10-e122-11e9-b112-9624ec9edc59.

115 Sanderson.

116 Deloitte, "2020 Renewable Energy Outlook," 2019, https://www2.deloitte.com/content/dam/Deloitte/us/Documents/energy-resources/us-2020-renewable-energy-industry-outlook.pdf.

117 Thomas L. Friedman, *Hot, Flat, and Crowded: Why We Need a Green Revolution—and How it Can Renew America* (New York: Farrar, Straus and Giroux, 2008), 241–242.

118 T. Wang, "Leading countries in installed renewable energy capacity worldwide in 2019," Statista, April 7, 2020, accessed June 23, 2020, https://www.statista.com/statistics/267233/renewable-energy-capacity-worldwide-by-country/#:~:text=Renewable%20energy%20capacity%20%2D%20country%20ranking%202019&text=The%20leading%20countries%20for%20installed,capacity%20of%20around%20264.5%20gigawatts.

119 United States Energy Information Administration, "How much of US energy consumption and electricity generation comes from renewable energy sources?," updated May 6, 2020, accessed June 24, 2020, https://www.eia.gov/tools/faqs/faq.php?id=92&t=4#:~:text=In%202019%2C%20renewable%20energy%20sources,about%2017%25%20of%20electricity%20generation.

120 Devin Hartman, "A Conservative Energy Reset," *National Review*, January 19, 2018, accessed June 24, 2020, https://www.nationalreview.com/2018/01/conservative-energy-policy-let-market-decide/.

121 Hartman.

122 Solar Energy Industries Association, "United States Surpasses 2 Million Solar Installations," May 9, 2019, accessed June 24, 2020, https://www.seia.org/news/united-states-surpasses-2-million-solar-installations#:~:text=Today%2C%20the%202%20million%20residential,will%20have%20a%20solar%20installation.

123 John F. Kennedy, Commencement Address at Amherst College, Amherst, MA, October 26, 1963.

124 Chris Martin and Millicent Dent, "How Nestle, Google and other businesses make money by going green," *Los Angeles Times*, September 20, 2019, accessed July 1, 2020, https://www.latimes.com/business/story/2019-09-20/how-businesses-profit-from-environmentalism.

125 Martin and Dent.